Advances in Spatial Science

The Regional Science Series

This series contains scientific studies focusing on spatial phenomena, utilising theoretical frameworks, analytical methods, and empirical procedures specifically designed for spatial analysis. Advances in Spatial Science brings together innovative spatial research utilising concepts, perspectives, and methods relevant to both basic science and policy making. The aim is to present advances in spatial science to an informed readership in universities, research organisations, and policy-making institutions throughout the world.

The type of material considered for publication in the series includes: Monographs of theoretical and applied research in spatial science; state-of-the-art volumes in areas of basic research; reports of innovative theories and methods in spatial science; tightly edited reports from specially organised research seminars. The series and the volumes published in it are indexed by Scopus.

More information about this series at http://www.springer.com/series/3302

Hugo Pinto • Teresa Noronha • Eric Vaz
Editors

Resilience and Regional Dynamics

An International Approach to a New Research Agenda

Springer

Editors
Hugo Pinto
Faculdade de Economia, Edificio 9
University of Algarve
Faro, Portugal

Teresa Noronha
Faculdade de Economia, Edf. 9
University of Algarve
Faro, Portugal

Eric Vaz
Geography and Environmental Studies
Ryerson University
Toronto, Ontario, Canada

ISSN 1430-9602 ISSN 2197-9375 (electronic)
Advances in Spatial Science
The Regional Science Series
ISBN 978-3-030-06981-0 ISBN 978-3-319-95135-5 (eBook)
https://doi.org/10.1007/978-3-319-95135-5

This Springer imprint is published by the registered company Springer Nature Switzerland AG
The registered company address is: Gewerbestrasse 11, 6330 Cham, Switzerland

Acknowledgements

The editors would like to thank all the contributing authors for their valuable contributions as well as Melanie MacDonald and Waleed Babary for proofreading the book. We would also like to thank the series editors for their support and positive comments. Finally, we would like to thank Barbara Fess and Marion Kreisel at Springer, Heidelberg, for their ongoing support and prompt replies to several queries.

Contents

Chapter 1
Introduction: Resilience—Concepts and Geography

Teresa de Noronha, Eric Vaz, and Hugo Pinto

1.1 The Emergence of the Concept

The notion of resilience emerges as a possibility to comprehend how different systems deal with shocks. It raises important issues regarding public policies and socioeconomic programs. Resilience has obtained a considerable degree of attention over the last years.

The concept recalls the vision of physics and natural sciences, referring to the stability of materials, its resistance to external shocks, and the capacity of the system of returning to the pre-shock state after suffering the impact. During the 1970s, Holling published an influential work in which he applied the concept of resilience to ecosystems, focusing on a system's ability to absorb shocks and retain structural functions. Holling understood resilience as a measure of the persistence of systems and the ability to absorb change and disturbance while maintaining relationships between populations or governmental variables.

Recently, however, resilience has become a more prominent term, mainly due to the financial and economic crisis, which has resulted in an understanding of the concept that goes beyond the strictly ecological aspects; resilience has gained a more socioeconomic character. In this last case, a robust system is one that can withstand strenuous system-shocks and re-stabilize itself, despite not precisely at the same equilibrium point as the status quo before. The problem with this definition is that, by considering merely equilibrium, resilience will always revolve around getting back

T. de Noronha (✉) · H. Pinto
Faculty of Economics, University of the Algarve, Faro, Portugal
e-mail: mtvaz@ualg.pt; hpinto@ualg.pt

E. Vaz
Department of Geography and Environmental Studies, Ryerson University, Toronto, ON, Canada
e-mail: evaz@ryerson.ca

H. Pinto et al. (eds.), *Resilience and Regional Dynamics*, Advances in Spatial Science, https://doi.org/10.1007/978-3-319-95135-5_1

to a specific trajectory or point without taking into consideration alternative pathways or, in other words: adaptation, change, and evolution. This is the reason why many authors do not rely on this meaning of resilience. They prefer to emphasize the evolutionary, multi-equilibrium perspective that allows systems to recover from shocks, not by only by going back to the previous states, but also by creating new alternatives.

The evolutionary approach to regional resilience focuses on the long-term capacity of regions to deal with shocks. It can be said that, in an evolutionary framework, resilience is not a mere property or goal, but rather an on-going process. Evolutionary resilience admits that often a regional economy, as a complex adaptive system, cannot return to the state it was before the shock in some instances. In such cases, a new point of equilibrium must be found, and preferably one that could be as efficient or, if possible, even more advantageous than the former. Inspired by this evolutionary perspective, an entirely new area of research on resilience begins to emerge in socioeconomic studies.

1.2 Pioneering Work on Resilience in Psychology

Most of the pioneering work related to social sciences, as highly complex systems, had its very origin on the more recent developments made by psychologists. In the scientific field, the concept reports a specific kind of strength originated by fragility and becomes a vibrant and promissory concept for social sciences in which marginal, deprived social contexts, struggling for survival, can find tools to grow and survive the competition.

To advance a research agenda for resilience and innovation, we first justify the goal of this book with so many contributions from sociologists and economists. A view over some facts related to the psychological approach of resilience are provided, and parallelism of those with the social context is conducted.

Norman Garmezy, a developmental psychologist, studied large samples of children for decades', observing how, in some cases, they excel regardless of the difficult circumstances they were exposed to. His primary goal was to focus on schools from economically depressed areas and to search for adaptive and successful children, despite difficult backgrounds.

Resilience offers a strange challenge: it is not an attribute evaluated on any test. It evolves as life reveals. If no adversity occurs, it will not be possible to detect how resilient one is. It is only when confronted with hindrances, obstacles or environmental threats that resilience, or the lack of it, emerges.

Along the several decades and studies related to psychological resilience, some direct causes for resilience have been identified: individual, psychological factors (reflecting some disposition and external) and environmental factors, which are more related to a random set of circumstances.

In 1982 and 2001, Emmy Werner discussed and published the results of a 30 year long project, allowing the identification of several factors of resilience out of the analysed data. These factors were related to chance and psychology: those more resilient detained an "internal locus of control" and trusted that only themselves could affect their achievements as creators of their opportunities. It was furthered that resilience was a dynamic concept, strongly temporal, and its changes over time nurtured the individual breaking points and skills, leading to unstable outcomes. These conclusions became beneficial results for social sciences in general, indicating not only that resilience is a dynamic concept, variable over multiple and strong stressors, but also that resilience can be acquired as a skill.

Later, Bonanno et al. (2007) emphasized adversity as to better formulate his theory of resilience utilizing the fundamentals of the stress-response system (a response to millions of years of animal evolution). If most of the people can make good use of that system to deal with stress, why do some use it more frequently or efficiently than others? As it seems, living through adversity, be it of endemic or of environmental nature, or through an acute negative event, does not guarantee that one will suffer going forward. What matters is whether that adversity becomes traumatizing or not.

Nonetheless, the most important remarks from the previous studies emphasize that positive attitudes can be taught by training people to regulate emotions in a long-lasting form better. In this sense, the research of Seligman (2011) shows that training people to change their explanatory styles from internal to external, from global to specific, and from permanent to impermanent made them more psychologically successful and less prone to depression. Also, the locus of control can change from external to internal, leading to positive changes in both psychological well-being and objective work performance.

To summarize, those cognitive skills that underpin resilience at individual levels can be learned over time, may create resilience where there was none, also working in the opposite direction, thereby threatening defense mechanisms and negatively affecting stability. From this short review over the roots of the concept of resilience we, as social scientists, must retain that there exists a set of skills that, although variable in time and space, can be identified and, if need be, integrated into an increasingly complex system—in our case, a social-economic system.

Facing a strong financial and economic crisis since 2008, it is not surprising that resilience rose to prominence in our field of research during the last decade. Social scientists go beyond a strictly ecological understanding of resiliency and define a resilient socio-economic system as a robust system that can withstand serious system-shocks and re-stabilize itself, albeit not exactly at the same equilibrium point as the status quo before.

Just like in the case of individual resilience in the psychological approach, the evolutionary approach to regional resilience focuses on the long-term capacity of regions to deal with shocks. A system's resilience depends not only on the capacity to recover, but to change, learn and prevent similar shocks in the future. Resilience is understood as the capacity to sustain long-term development and to respond positively to short-term shocks—it is an on-going process.

This very simplistic approach reveals the major goals of the present publication: To propose to intertwine resilience and innovation and to further design a research agenda covering the following major topics:

- Theoretical contributions towards the integration of resilience, innovation, and regional science
- Empirical studies focusing the conditions for resilient territories
- Smart specialization and innovation
- Impacts of resilience in regional development
- Clustering dynamics, and resilience
- Comparative studies on institutional factors that shape resilience
- Policies implemented in resilient territories

1.3 Towards the Geographies of Resilience: The Case of Europe

Europe has been facing a long-standing crisis over the last years. Beginning in 2007, this crisis has been described as the worst economic crisis since the Great Depression of the 1930s. Based on Eurostat data regarding the change of GDP, unemployment rate, and R&D expenditure at NUTS 2 level, and comparing the performances across EU regions, we can confirm that, in terms of GDP, between 2008 and 2013, Southern European regions and the United Kingdom registered the most negative variations. Netherlands, Finland, Greece, Italy, Cyprus, and Spain faced the worst situations. The interior regions of France and some parts of Sweden also had a negative economic growth, although to a lesser extent than the cases mentioned before. Conversely, the regions of Eastern Europe, Denmark, Germany, Belgium, and some French regions had grown. The most noticeable pattern visible is that the regions of southern Europe suffered more, even if some countries such as the United Kingdom and Finland have had a negative performance as well.

Further conclusions based on the data show the changes to the unemployment rate, between 2009 and 2014, similar to some extent to those verified with the GPD changes. The regions from Southern Europe remain the ones in the worst situation, the case of Portugal, Greece, Cyprus, and, to a lesser extent, Italy. France has mixed results, with the central regions having the worst performances. Netherlands and Bulgaria are close to Mediterranean countries in terms of performances, and Poland has mixed results. United Kingdom has different results regarding unemployment comparing with GDP. Overall, the northern and central European regions made considerably better than other regions. If we consider the changes in investment in R&D between 2007 and 2012, most EU regions did not perform cuts on the domestic expenses on research and development during the selected period.

The only countries with regions that registered cuts in R&D were the United Kingdom, Portugal, France, Sweden and Finland, with the latter being the only country to register cuts in all its regions. Besides the particular regions of the aforementioned countries, many regions in the EU increased their expenditure in R&D, with the regions of Central and Eastern Europe showing the biggest increases. It is also worth mentioning that, while there is a degree of contrast between Southern and Northern Europe, in the sense that the Southern countries invested less than the Northern ones, this gap is relatively smaller than in both previous cases.

As we illustrate the different regional capacities of coping with the crisis, we must confirm the importance of resilience to surmount difficulties and better manage shortages in capital availabilities. These maps visualize the reality that some countries and groups of countries were/have been more resilient regarding the adaptation of production and employment to the crisis than others. The fact that there is not a clear pattern regarding R&D change may induce the conclusion that many different factors besides R&D may be able to determine the resilience capacity of regions and countries. For example, in the conclusion of this book, we observe the surprising revival of the Portuguese economy, whose GDP growth rate grew, against all the expectations, from −4%, in 2012, to +1.6%, in 2015.

1.4 Organization of the Book

This book collected a group of contributions from the fields of Regional Science, Economics of Innovation, Science and Technology Studies, and Planning, thus comprehending a multidisciplinary approach in both theoretical and methodological contexts.

So far, few books have dedicated specific attention to the intersection between innovation and resilience. In this case, we have organized this publication into five major parts:

Chapter 1—The introduction of the book Hugo Pinto and Teresa Noronha, revisit the pilot concepts of resilience mainly from a socioeconomic and psychological perspective. Thereby, they supply a bridging understanding into its most recent use by the social sciences. A discussion on the emergency of the concepts related to resilience is supplied as much as the variety of forms to observe it is explored. Further, maps are used to ullustrate some of the variables closely related to resilience in Europe.

Part I of the book entitled Theoretical Foundations comprehends a theoretical framing on which most of the chapter relies. This part is very useful to consolidate the still embryonic concept of resilience when applied to Social Sciences.

Chapter 2 "Evolutionary Complexity Geography and the Future of Regional Innovation and Growth Policies" authored by Philip Cooke, emphasizes the complexity and how the evolutionary perspective may shift the scope of regional innovation and growth. By reviewing some key conceptual and practical barriers that have hampered territorial, economic development prospects in most advanced countries for some time, the author searches for a great escape from cognitive and policy "lock-in" situations. A brief review of evolutionary economic geography (EEG), refashioned as evolutionary complexity geography (ECG) is made.

Chapter 3 "Evolutionary Resilience Shifting Territorial Development Paradigms" by Carlos Gonçalves continues adds foundational understanding to concepts of evolutionary resilience demonstrating the recurrence of crises in the contemporary societies and conferring centrality to evolutionary resilience, also by resuming pre-crisis trajectory.

Part II refers to the Multilevel Aspects of Resilience and includes a set of four chapters highlighting empirical findings from an extensive international basis and helping claim for the multi-level character of resilience. This we understand to be the most pioneering aspect of our research agenda.

Chapter 4 "Economic Crisis, Turbulence and the Resilience of Innovation: Insights from the Atlantic Maritime Cluster" by Hugo Pinto, Elvira Uyarra, Mercedes Bleda and Helena Almeida suggests the notion of 'resilience of innovation' as the capacity of an innovation process to maintain its function at different levels of operation.

In Chap. 5 "Innovation, Regions and Employment Resilience in Sweden" Charlie Karlsson and Philippe Rouchy adopt a mixed approach of resilience associating economic geography with labour capital, applying it to a sample of Swedish regions from the perspective of labour accessibility, performance, and dynamics.

In Chap. 6 "Diversifying Mediterranean Tourism as a Strategy of Regional Resilience Enhancement", André Samora-Arvela, Eric Vaz, João Ferrão, Jorge Ferreira, and Thomas Panagopoulos assess local resilience for the tourism industry, addressing future challenges of the industry in regards to environmental degradation and climate change.

Part III focuses on a possible research trend Towards Strategies for Resilience that emerges from the three sequential chapters:

Davide Fassi and Carla Sedini focus, in Chap. 7 "Design Solutions for Resilience" on the methods and tools which are typical to design processes and put in place strategies to improve the quality of life of neighbourhoods. Many intangible elements do well with social capital and therefore must be considered to design methods that are very useful in highlighting and explaining some social contexts. Narratives, participation, co-design are suitable approaches to create and make visible connections which can help resilience to be improved. Such design tools do not only ease the success of projects but also involve institutions and citizens.

Sandro Guduchi and Manuel Fernández-Esquinas Chap. 8 "Organisational innovations for science-industry interactions: the emergence of collaborative research centres in Spanish regional innovation systems" suggest that the creation and diffusion of organizational innovations for knowledge transfer in regional innovation systems is a phenomenon relevant to the understanding the innovative dynamics associated with regional resilience. They argue that the collaboration between the scientific and industrial sectors is of critical relevance. This chapter analyses the specific case of organizational innovations of the collaborative research centres (CRCs) in Spain constituted by the joint participation of scientific, government, and industrial agents to increase the technical, scientific capacities of the territory and guiding them towards strategic sectors for the development of the productive system.

Within the same frame of understanding, in Chap. 9 "Merging entropy in self-organization: a geographical approach" Eric Vaz and Dragos Bandur propose the integration of innovative spatial analytical methods to measure self-organization spatial phenomena at the regional level, suggesting that machine learning may have a significant impact on using concepts of entropy for novel techniques of clustering and classification, important assets to support regional decision making.

The next and final, Part IV, Resilience, and Innovation fine tunes the concept of resilience while intertwining it with innovation, technological or organizational, and bringing to light a new conceptual basis of increasing resilience based upon new attitudes, materials, strategies.

Chapter 10 "Innovative Urban Paradigms for Sustainability and Resilience" by Manuela Pires Rosa, supplies an extensive set of social and cultural advances which imply deep changes in urban paradigms, namely in the areas of land use, transportation and water planning and management. She explains how some innovative approaches in the field may contribute to reduce the vulnerability of the territories, to conserve natural resources, and to avoid environmental pollution, making them able to develop collaborative and adaptive management processes that are therefore more resilient.

In Chap. 11 "Innovation as Transformation: Integrating the Socio-Ecological Perspectives of Resilience and Sustainability" Karl Bruckmeier and Iva Miranda Pires connect the concepts of social and ecological resilience and sustainability to develop an integrated perspective of innovation. They discuss possible combinations of resilience and sustainability concerning innovation, adaptation, and transformation for local strategies for urban-rural development in metropolitan areas and their surroundings. The analysis is illustrated with studies of agricultural development in peri-urban and urban areas where the interaction of urban and rural development increases the difficulties of transformation to sustainability.

Finally, as a contribution to this conceptually new approach, in Chap. 12 "Territorial Innovation Models: Which Consequences regarding Policy Design for Peripheral Regions? A Portuguese Perspective" Domingos Santos's chapter main objectives are to discuss those theoretical frameworks that may enable a better understanding of the relationship innovation-territory, analyzing the characteristics of the Portuguese context and, finally, the main implications in the design and implementation of territorially embedded innovation.

1.5 Final Remarks

As we know, several countries across the world and some European regions have experienced a rise in unemployment and reduction of economic growth suffered as a result of the economic crisis. Such problems seem to subsist, which negatively impacts some regions more than others, damaging the normal economic turbulence, the innovative dynamics of the territories, and the consequent asymmetric effects upon European regions. Thus, investigating how resilient the innovation process to crises has been is worthy of attention. This research helps us better understand the capacity of the innovation systems to resist and adapt to disruptions such as the ones generated by economic downturns.

As seen, resilience encompasses many dimensions, which often transcend the simple measures of product or employment growth. Research and innovation strategies (RIS3) for smart specialization, as well as focused S&T and innovation may make a significant contribution towards more resilient territories in the European regional policy context.

This book discusses how innovation may help to configure a new regional institutional framework oriented to a smart specialization that takes into account the resilience of territories and regions. At the same time, it also analyses methods emphasizing that institutional diversity and integration may promote more business discoveries, spillovers and agglomeration efficiencies that not only ease structural change in regions but also help to construct resilience.

Working to establish a new, more complete definition of resilience and innovation will bring a positive contribution to regional science overall, but this body of work is particularly addressed to policymakers, to regional authorities dealing with regional governance and development strategies, and to those responsible for risk management, at multiple levels of action, helping the construction of more resilient territories in Europe and across the world. Public policies and the European structural and investment funds should pay more attention to the structural weakness of certain regions.

References

Bonanno, G. A., Galea, S., Bucciarelli, A., & Vlahov, D. (2007). What predicts psychological resilience after disaster? The role of demographics, resources, and life stress. *Journal of Consulting and Clinical Psychology, 75*(5), 671.

Seligman, M. E. P. (2011). Building resilience. *Harvard Business Review, 89*(4), 100–106.

Part I
Theoretical Foundations

Chapter 2
Evolutionary Complexity Geography and the Future of Regional Innovation and Growth Policies

Philip Cooke

2.1 Introduction

> By its nature, the metropolis provides what otherwise could be given only by travelling; namely, the strange (quoted in Jane Jacobs 1961, 238).

The key idea tested here is that territorial (meaning urban, regional, national or international scale) knowledge flows have changed under knowledge economy conditions. Does knowledge still flow sectorally in specific industries in the main? Are multinationals still dictating knowledge flows in supply chains? Is policy-maker attachment to 'specialisation' of economic development in vertical 'knowledge silos' such as clusters appropriate? Surprisingly, perhaps, the answers to these and related research questions after 5 years of recent ECG research into Regional Innovation Systems (RIS) were largely in the negative, although as innovation theory shows, every paradigm shift meets initial resistance from the ancien regime. Innovation of the systemic variety is the main reason why knowledge dynamics have become less vertical, cumulative and path dependent and more transversal, combinative and path creating. Systemic innovation is linked through networks of buyers and suppliers of knowledge, goods and services (so-called "entrepreneurial ecosystems"). In passing, this gives an answer to a sometimes asked question, which is what, exactly, is innovation for? The purpose of innovation is growth, measured in terms of productivity, efficiency and effectiveness. It seems that capitalism, which from a Schumpeterian perspective is fuelled by innovation, must grow in order to survive. Growth is implicit in the notion of markets, the inefficiencies in which stimulate innovative efforts to profit from seeking better alignments between value

P. Cooke (✉)
Mohn Centre for Innovation & Regional Development, Western Norway University of Applied Sciences, Bergen, Norway
e-mail: CookePN@cardiff.ac.uk

H. Pinto et al. (eds.), *Resilience and Regional Dynamics*, Advances in Spatial Science, https://doi.org/10.1007/978-3-319-95135-5_2

and price,[1] whether of commodities, companies or currencies. For more citizens to have access to the quality of life of the typical middle-class household of the advanced economies is not a morally indefensible position, given the massive inequalities that arise from neoliberal dogma in many such countries, let alone between them and the developing world.

Growth is increasingly sought and found by firms and relevant support organizations exploring 'relatedness' within and beyond regional territories. 'Relatedness' means firms that understand each other's business models, skillsets and technologies even though they are in different industries. Hidden in different industries, firms may nevertheless offer innovative learning opportunities if they can be identified. This perspective is supported by at least three new territorial models. The first is that of New Economic Geography (NEG) which saw systemic regional innovation in terms of labour pooling behaviour. This means that firms and workers seek out regional market-size and financial spillover effects, co-locating or agglomerating when they find a region where industry has a lead over everywhere else, innovation being the explanation for that lead (Felsenstein 2011; Krugman 1991). Some modelling deficiencies persist in this perspective since it continues to produce misleadingly over-specialised and over-concentrated spatial results.[2] An alternative that does not fall into the trap of over-emphasising a single type of knowledge determinant of regional growth is New Growth Theory (NGT) with its better insights into endogenous (i.e. local or regional) technological growth. Here, by analysing regional knowledge externalities or spillovers interactively with human capital or labour mobility, the approach estimates the way that human and physical capital, labour mobility and innovation impact on regional productivity and growth (Martin and Sunley 2006). Following NEG theory for a moment, increasing returns theory also supports the deduction that the higher the average level of human capital, the more rapid the diffusion of knowledge, therefore the higher the level of regional productivity (including earnings; Felsenstein 2011). So NGT allows different kinds of regional knowledge and innovation into the innovation-productivity analysis. However, while human and physical capital combine positively to affect regional

---

[1]An anonymous referee queries this distinction. It is hoped that the following illustration is helpful. The price to a plumber to fix a burst pipe at a customer's home may be €5 for travel, materials costing €2.50 and an hour's labour at €10. However, the value of the service to the customer—who may have water leaking all over their house—is far greater than that, so the plumber typically estimates the price the customer will pay at €100. Investment bankers 'arbitrage' such 'value to price' differences for profit in the financial services industry.

[2]Krugman (1991) displays the centrality of innovation in his theory of city agglomeration while admitting it is simplistic: "There are assumed to be two technologies for producing manufactured goods: a "traditional" technique that produces goods under constant returns at a unit cost c1, and a "modern" technique with a marginal cost lower than c1, but that involves a fixed cost F per production site... If manufacturing is dispersed, an optimally located modern plant will be a distance of 1/4 from its average consumer, and will thus incur transport costs tx/4. On the other hand, if all manufacturing were concentrated at $z = 0.5$, an urban plant located at the same point could serve a fraction π of consumers at zero transport cost, and incur transport costs of only $(1—\pi)$ tx/4..." This story bears an obvious resemblance to the Big Push story of Rosenstein-Rodan (1943).

productivity, the model results are weakened by a "regional innovation" effect. Thus a third approach receives some degree of support from this inconsistency, namely Evolutionary Economic Geography (EEG). This perspective sees institutions, organizations and cultural practices as critical in generating regional growth. Thus cultural and institutional proximity are as important as spatial proximity and the region represents an active innovation agent. It has recently been termed Territorial Embeddedness Innovation (TEI) in contrast to exploratory "scientific and technological innovation" (STI), on the one hand, and "doing, using and interacting" (DUI) knowledge exploitation for innovation, on the other (March 1991; Nunes and Lopes 2015; Jensen et al. 2007).

Accordingly, this contribution summarises new arguments and findings for territorial knowledge dynamics that pose problems for prevailing knowledge about innovation and knowledge theory. The chapter is constructed around answers to four such problems, raised by the testing of a core ECG-informed theory supported by wide-ranging and structured evidence. This approach is marked by two sub-sections: the first is theoretical; the second is empirical. The first of the theoretical questions is: does the interactive model of innovation that replaced the prevailing linear model now itself require re-engineering? The linear model proposed innovation followed a line from R&D to prototyping and testing to commercial innovation on the market, the interactive model proposed feedback among suppliers (entrepreneurial ecosystems) in value chains. The second theoretical question, deriving from the Schumpeterian heritage, is what counts as 'radical' innovation? Does it only occur once every 60 years (long wave theory proposes "mechanisation" in railways during the nineteenth century being radically overhauled by "electrification" and "automatisation" in motors and vehicles in the 1900s and "informatisation" in computers in the late twentieth, early twenty-first centuries) with associated regulatory regime resistance (sometimes stimulus) lasting over lifetimes, or does it occur more frequently? This means involving swifter paradigm (economic drivers) and regime (regulation by government) change, especially in some industries or industry platforms displaying relatedness? Next, more practically, are innovators also entrepreneurs or do the complexities of distributed knowledge dynamics mean there is a diversity of (global) actors helping the translation of knowledge into commercial products and services? Does the new knowledge dynamics thinking make path dependence (historical industrial development trajectories) redundant, or is knowledge used for 'branching' and new path creation where transversal (crossover) knowledge dynamics are exploited? These issues will be addressed and their resolution illuminated by reference to ECG research findings (after e.g. Frenken 2006).[3]

---

[3]An anonymous reviewer holds that EEG after the Dutch approach should be cautioned against because it suffers from "ergodicity" i.e. all future states of the model must be in the model at the beginning. A priori, this seems unlikely for any kind of economic geographer given that in Boltzman's initial formulation the term refers to a '...dynamical system which, broadly speaking, has the same behaviour averaged over time as averaged over space.' Moreover, EEG research shows that 'relatedness' which equates very much to thinking on territorial knowledge dynamics (TKDs) includes 'revealed related variety' unpredictable ex ante but rather only understandable ex post.

2.2 Evolutionary Economic Geography Theory

This section will say little about NEG or NGT but much more about EEG (Boschma and Martin 2010). Evolutionary economic geography theory is a good example in itself of the evolutionary biology concept of 'exaptation' (Vrba and Gould 1982). The late evolutionary biologist Stephen Jay Gould held that a new word was needed to account for the biological process whereby an obsolescent organ evolved a new use over the long term, possibly even in a different species. Examples included the fact that human inner ear bones were once the jawbone joints of an extinct fish species and fish with buoyancy bladders have exapted the lung functions of earlier amphibious species, so the word proved useful. Evolutionary economic geography is a new discipline which has exapted concepts as old as nineteenth century classical economics, the forebear of the neoclassical perspective. "Cumulative change" Veblen's (1898) precursor of Myrdal's (1957) "circular cumulative causation" (CCC) was an early species of "increasing returns" (Krugman 1995). New neoclassicals created NEG by relaxing neoclassical assumptions including "constant returns", "perfect information" and "equilibrium outcomes". Evolutionists are as interested in increasing returns, appropriated by "new neoclassicals" like Krugman, for understanding basic spatial growth processes as neoclassicals are. But that interest is far less mechanistic and reductionist, emphasising much more the institutional, co-evolutionary and path dependent (historical) aspects of change (Martin and Sunley 2010). EEG also favours disequilibrium rather than equilibrium or even partial-equilibrium explanations for the crisis-ridden "progress" of capitalism. It does not assume economic balance and stability are normal but rather the reverse, namely that they are unusual and economic crisis conditions reflect such general conditions of instability.

The co-evolution of (regional) institutional regimes and related (regional) paradigms or economic mixes of industries is an extremely fruitful way to conceive of regionally adaptive (changeable) systems of innovation. This is because in explaining innovation and growth it is as inadequate to privilege "external shocks" as it is to privilege "endogeneity" (internally-generated growth impulses). If we think of regional regimes as regionally varying combinations of organisational or governance structures interfacing with institutional conventions (rules of the game), we immediately have a conceptual grasp on regional variety. This combination of formal governance or regulatory rules and informal practices of, for example, business associations, indicates an important source of regionally distinctive outcomes. We can think of these in terms of hierarchical adaptive system interactions. Thus everywhere is—in economy, politics and culture—different rather than identical because regions (and nations) vary within systems such as multi-level governance (e.g. the system involving the EU, member-states and regions). If to that is added the notion of regional paradigms as related varieties of path dependent "socio-technical systems" (industry mixes that comprise the regional or a national economy; Geels 2007) it is the interaction of these knowledge flows that produces innovation. Arthur (2009) calls this "combinative (or combinatorial) evolution" in his book on the nature of technology and innovation. For Martin (2010) this constitutes "path interdependence" a far more dynamic concept than "path dependence" because it is in such—what we term "recombinant" knowledge "collisions"—that all innovation lies (Schumpeter 1934). So we move from a vertical, linear and sectoral view of knowledge flows to one that recognises a more "geographically informed" epistemology of horizontal, interactive and inter-sectoral knowledge flows for innovation.

These are bold claims that require further elaboration. Put simply, Arthur's most recent statement about the ubiquity of 'bricolage' (recombination) as the midwife of all innovation may, from some perspectives—underestimate the role of truly novel knowledge. However, in engineering, which was Arthur's first calling and from which he gets much exemplification (e.g. the complex path dependence of jet engine technology), it is probably a more reasonable assertion than in, say, biotechnology, which he also declaims upon. Even some keystone biotechnology knowledge like DNA nevertheless betrays a "ghost in the machine" of metaphors exapted (borrowed) from elsewhere, such as the physicist Schrödinger's idea that DNA might resemble a non-repeating crystal. So what constitutes truly novel knowledge? Briefly, two examples must suffice: the first was the 2000 Nobel Prize-winning research by Heeger et al. (1978) which revealed the prevailing scientific consensus that polymers could only insulate not conduct electricity to be wrong. It is now the basis for Samsung's Active Matrix Organic Light Emitting Diode (AMOLED) technology which replaced liquid crystal in the screens of its Android and 4G LTE smartphones. The other is the nanotechnology research of Maria Strømme and her team at Uppsala University (Nystrom et al. 2009) on the filtering properties of special paper which, when trialled in a lake suffering eutrophication (algal blooms and de-oxygenation) produced electrolytic effects from the interaction of the filter paper with specific algae. A method of utilising algae to store electricity in a battery was thus discovered from a completely unknown source. The battery can be recharged much faster than a lithium battery. The cellulose that Strømme and her colleagues use comes from a polluting type of algae whose cell walls contain cells with a distinctive nanostructure, which gives it 100 times the normal surface area. The researchers coat paper made from this cellulose with a conducting polymer then sandwich a salt-solution-soaked filter paper between the paper electrodes. It charges in a few seconds, is flexible, sustainable and non-toxic. Hence though the battery application utilises the conducting polymer, the discovery represents novel knowledge about the electrical storage capabilities of algae, possibly a solution to the age-old problem that electricity is hard to store at scale and over lengthy time-periods. Of course, nature already possessed these properties, so they were not strictly "new", which tells us something about the ontology of knowledge, scientific or otherwise. Accordingly, we conclude this "nothing new under the sun" debate by asserting that the novelty of innovation lies in its recombinations rather than its ingredients, which were always there in atomic, molecular or memetic forms, awaiting discovery. It should be noted that algae contain many previously undiscovered yet potential commercial opportunities, for example synthesis of Omega-3 nutrients from rapeseed oil.

In complexity theory, these knowledge and innovation processes would be referred to as exploration of the "adjacent possible" in the first case, and "preadaptation" rather than the more biological "exaptation" in the second. The adjacent possible is a search process that seeks novel solutions, many being incremental innovations, relatively close to the existing state of the art. Such novelty becomes radical innovation when the knowledge recombination search swiftly reveals numerous related innovation possibilities and potentials. In the case of paper batteries, the adjacent possible was the application of old knowledge (conductive polymers) to new (electrolytic algae) to create an eco-innovation. Preadaptation, which is a more common kind of innovation process, starts with already existing innovation which is then preadapted to a new setting either by some kind of "cognitive reversal" (retro-innovation; Dew et al. 2004; Villani and Ansaloni 2010) or by adaptively transferring it from one industry into a wholly different one (Kauffman 2008). Kauffman's exemplar of cognitive reversal preadaptation concerns the invention of the modern tractor, the early massive engines for which continually broke the chassis when mounted. An engineer, noting the scale and rigidity of the engine block, suggested it could form the chassis. The historical innovation was Henry Ford's Fordson Model F which was completed in 1916 and was the first lightweight, mass produced tractor in the world. Ford engineer Eugene Farkas successfully designed the engine block, transmission, and axle housings bolted together to form the basic structure of the tractor. By eliminating the need for a heavy separate chassis, costs were reduced and manufacturing was simplified. We could point to the Wright brothers' innovation of the aeroplane which combined bicycle, boat, kite and early combustion technology in the form of wheels, chains, propellers and motors from different industries—all combined to fulfil their purpose, namely to create a flying machine.

Today, preadaptation is consciously practised by regional cross-cluster/sectoral knowledge transfer agency Bayern Innovativ (BI) for its industry members. This involves large numbers of variably-sized, themed meetings of industry innovators evaluating the preadaptation (or knowledge/innovation transfer) potential of innovations already implemented in other industries, as described in Cooke et al. (2010). In this book one good example was the occasion when BMW was exhibiting the nanotechnology-refined textile that kept the seats of its new model free from attracting dirt. Nano-filters had been embedded in the seat fabric to produce this effect. Sitting in the audience were representatives of hospitals and medical clinics. They immediately thought that such an innovation might reduce the bad effects of bacteria and dirt sticking to medical uniforms if a suitable textile could be produced with the same filtering properties. Over time such "innovation-transfer" was achieved and the new product is on the market. So much innovation—in the form of commercialised recombinations—has occurred historically that transversality across them will typify innovation opportunities in the future. Currently, transfer occurs face-to-face and by word-of-mouth but it is easy to see how a firm or agency could make such knowledge available as a digital ("app") market offer.

2.3 Territorial Knowledge Flows and Innovation Issues

2.3.1 Does the Interactive Model of Innovation that Replaced the Prevailing Linear Model Now Itself Require Re-engineering?

Thus it may be deduced that the conventional wisdom about innovation is in need of an overhaul. It was noted at the outset that transversal or 'crossover' knowledge flows not only pose problems for the cumulative model of innovation but also for the linear (STI) and interactive (DUI) versions of this model that have dominated understanding of innovation for decades (Balconi et al. 2010; Kline and Rosenberg 1986). Both share verticality, the first from its emphasis on intra-corporate knowledge flows from R&D laboratory to marketing and sales departments, the second from recognition that supply chains were more clearly emergent with the onset of Japanese modes of "lean production". The older focus upon innovation without much thought either of what it was for, or how knowledge acquisition to achieve innovation was related to it, means one of two things. First, it could be that innovation was once linear, cumulative and closed but that is no longer the case. This seems unlikely from a complexity perspective because both Kauffman (2008) and Arthur (2009) stress that the key feature of complex adaptive socio-economic systems is that:

> ...The more diverse the economic web, the easier is the creation of still further novelty... [leading to]...a positive correlation between economic diversity and growth (Kauffman 2008, 151–160)

and, as Arthur sees it:

> ...When a network consists of thousands of separate interacting parts and the environment changes rapidly, it becomes almost impossible to design top-down in any reliable way. Therefore, increasingly, networks are being designed to 'learn' from experience which simple rules of configuration operate best within different environments (Arthur 2009, 207)

What is more likely is that the "framing" of these innovation models was wrong. This means observers misunderstood and then over-simplified what they thought they had seen, or perhaps not seen because most innovation occurs in confidential situations. Contrariwise, what was always present even in portrayals of intra-corporate or intra-supply chain innovation orderliness was a great mixture of purchasing or 'borrowing' of adjacent extra-mural ideas, possibilities and solutions from related and even unrelated industries. Individual scientists, knowledge entrepreneurs and consultant experts come to mind as innovation contributors here. Even Alexander Fleming, who innovated antibiotics, was helped by his housekeeper in noticing his discovery of penicillin which she thought was cheese. Accordingly, other than describing such 'bricolage,' (recombination) theorists at the time lacked an interest, or a theoretical discourse, within which to position such messy processes. So, second, evolution of knowledge flows around 'platforms of innovation' integrated nowadays by digitisation as facilitators of economic growth. This has both shattered the hitherto prevailing narrative of cumulative orderliness and introduced "an image of wholeness, and within that wholeness a 'messy vitality'" (Arthur 2009, 213).

2.3.2 *What Counts as 'Radical' Innovation?*

If all innovation is 'bricolage' (recombination) where one innovation builds upon a preceding one or more to fill a niche created by an opportunity arising from what has gone before (in ECT/ECG "self-organising systems"), it seems hard to find a place for other than incremental innovations that explore possibilities of preadaptation or the adjacent possible. Kauffman (2008) frequently uses the tractor metaphor to marvel at the ingenuity of mankind but he also notes how, for example, the innovation of the remote TV channel control could simply not have been envisaged or imagined in a society without TV, or more particularly, multi-channel TV. This gives a clue to the reasons why it is important to differentiate between innovation in general, most of which is preadaptation or adjacency, therefore mostly incremental, and radical innovation. Whether that means most innovation occurs in geographic proximity is an open question to which we will return. But, for the moment, research on the history of innovation (e.g. Johnson 2010a, b) suggests most of it is produced in geographic proximity to where 'adjacent possible' opportunities arise, much of it contains unexpected elements (e.g. paper research finds electrolytic algae), and even if knowledge flow interactions are inter-continentally relational, innovation is recombined at the spatial point of the innovator (team). Johnson (2010a, b) accepts only one exception to this rule, namely the "multiple" as when an innovation (e.g. the incandescent light bulb) occurs simultaneously and independently in different regions. Hughes (1977) argues Edison gained priority for this because he also innovated a co-evolving electricity generating and lighting system. This is a clue to the difference between long-term radical innovation and short-term incremental innovation since the former swiftly stimulates a variety of related innovations. Time and variety distinguish systemic from routine innovation, rendering the first "epochal" in the sense of ushering in a long-wave technological regime enveloping, protecting and facilitating exploitation of the new growth-inducing technological paradigm—classically as with our contemporary informational economy.[4] But within that technological paradigm many shorter-term but still radical innovation "episodes" occur, today affecting retail, newsprint, recorded music and even taxi transportation firms. Finally, time is important in giving rise to more "episodic" radical innovation. Thus change occurs more swiftly in design, creative and "cognitive-culturally" inspired industries like smartphones than in light bulbs. Here instant shifts in socio-cultural

[4]Thus a smartphone has a QWERTY keyboard, invented by Sholes of Milwaukee in 1878, a camera (Fox Talbot 1841; digital version Kodak 1975), a sound system (Edison 1877; digital version Alec Reeves 1938), Internet and email via packet-switching (Donald Davies 1965; ARPA/Vint Cerf/Robert Kahn 1969); social media on Web 2.0 (1999); "apps" innovated by Apple AppStore in 2008 and so on.

meaning can be captured through the phenomenon of "circles" in design driven industries or "crowdsourcing" and "crowdfunding" as practised by "apps" firms in the smartphone industry (Scott 2008; Pisano and Verganti 2008; Page 2007). So, we conclude that the original idea of radical innovation survives but needs variegation conditional on different temporal innovation "frames" whose knowledge turnaround speeds are conditional on their conscious exploitation of "crossover of knowledge or actual innovations among firms or industries)—'transversality'" (Cooke 2013). Illustrative material on this for Sweden's regions of Skåne and Västra Götaland and the French Midi-Pyrénées is presented below.

2.3.3 Are Innovators Also Entrepreneurs?

This question addresses the complexities of distributed knowledge dynamics asking do they mean there is a diversity of (global) actors assisting the translation of knowledge into commercial products and services? This is not the old individualist question about believing innovation to be the product of genius. It is far more important than that and relates to a common misconception that entrepreneurship and innovation are different sides of the same coin, or worse, that they are the same thing. If it was ever true, it seems decreasingly so nowadays. Even Schumpeter (1934) is pretty clear that the key skills were very different because the innovator recombined knowledge while the entrepreneur assembled the financial, legal and human resources to commercialise it. EEG research has registered the rise of complexity in the intermediation of innovation processes by practitioners of knowledge-intensive business services who are found performing crucial co-ordinating, advisory and consulting roles in most industries (knowledge intensive business services—KIBS—Strambach 2010). These include entrepreneurial ecosystems of management accountants, venture capitalists, patent lawyers and so on. Even knowledge-intensive business services (KIBS) for farming are located in cities where insurance, credit and technical talent is found rather than within the rural markets for such services. But KIBS are themselves a very large platform of differentiated knowledge. This returns us, momentarily, to the question of geographic proximity raised above. Clearly, the phenomenon of rural services being supplied from metropolitan locations reveals how the presence of global talent pools, their knowledge spillovers and "relatedness" across industry boundaries allows for fluid entrepreneurial activity to be conducted in an urban ecosystem by KIBS businesses of a multitude of sizes. Ironically, indicators of such knowledge-intensive entrepreneurial concentrations that place cities like Stockholm and London at the peak of the European hierarchy for their disproportionate shares of employment in KIBS and the lesser category of high-tech manufacturing (e.g. Cooke and Schwartz 2008) also show London, at least, to underperform UK regions on innovation per capita (Chapain et al. 2010). So it seems likely that knowledge-intensive entrepreneurs are located in a different place from innovators. More precisely, most KIBS and high-tech manufacturing workers in cities are clearly neither entrepreneurs nor innovators. Rather they are clerical, secretarial, retail and administrative workers, a corrective to the discourse that emphasises the creativity of large cities, at least regarding the composition of their labour markets. So we conclude from research like that above on cities that entrepreneurs are increasingly divorced as actors and increasingly so in geographical terms. This is a source of the difficulty innovators have in launching new start-up businesses, especially in Europe.

2.3.4 Does the New Knowledge Dynamics Paradigm Make Path Dependence Redundant?

This is possibly the most interesting question of all posed by EEG and ECG research. Traditionally, path dependence has been associated with somewhat negative outcomes like "lock-in" of older industrial regions to outdated industry and management practices (Grabher 1993). David's (1985) equilibrium perspective overemphasised such "lock-in" issues. Nowadays it is criticised in favour of a more open and innovation-friendly perspective (Martin 2010). A second weakness was Arthur's (1994) reliance on "chance" or "accidental" explanations for innovative events that shift path dependence (Martin and Sunley 2010).

Building on a more socially constructive conception of path dependence, reflective of Garud and Karnøe's (2001) notion of innovation also involving "mindful deviation" by social agency to effect change, EEG has introduced the notion of path inter-dependence. Martin and Sunley (2010) thus align this adjusted perspective on path dependence to another key EEG concept, namely "proximity" to move closer towards our mobilising explanation for innovation which, when linked to the multi-level perspective (MLP) idea of co-evolving socio-technical systems (STS; after Geels 2007) allows us to incorporate the key ECG concepts of "preadaptation" and the "adjacent possible" in a rather satisfactory explanation of emergent regional knowledge flows and innovation. Allowing for the likelihood of market failure by firms to explore regional paradigm (regional economic mix of industries) relatedness sufficiently, delaying the onset of new path creation, this also opens up a key regional regime (government/governance) opportunity.

This is to introduce firms to both regional and non-regional innovation as a "preadaptive" form of transversality and to encourage exploration of "structural holes" or "white spaces" among regional paradigm (the economy's socio-technical system) elements (Johnson 2010a, b). Thus we begin to see more clearly the element of "path inter-dependence" that defines key spatial forces underlying and influencing inter-organisational relations. Martin and Sunley (2010) mean it largely in terms of the economic geography (paradigm) dimension, including inter-dependent technological paradigm interaction. This will be explored in more detail under the rubric of "relatedness" conjoined to "transversality". Accordingly, paradigm "relatedness" may be latent but a regional regime may realise and reveal it for exploitation. It can be done by a BI-type "living laboratory" or by means of the appropriate "smartphone app" when it exists. This is a key regional innovation policy breakthrough in regional regime/paradigm interaction because 'transversality' is the policy correlate of relatedness among industries or firms. Policy—whether by government, public-private governance, or private governance by intermediary or lead-firm initiative—may be active where market failure means that potentially complementary firms or industries in geographical proximity never meet to discuss possible innovations. If policy is not active, then innovative 'structural holes' will remain unidentified unless and until firm "search" of the selection environment eventuates, possibly due to the rise or entry of new incumbents (see below). High market uncertainty in a context that values "innovation" as the highest virtue of the accomplished firm (and region) owing to its overwhelming contribution to productivity and growth, means regional regimes (or governance systems) will increasingly assist such search for structural holes by inducing speed-up in the process.

2.4 Empirical Tests of the Foregoing: Brief Comparative Case Analysis

2.4.1 Region Skåne

ECG research shows strength in this region in Sweden to be clustered in agro-food production and services, including functional food based on biotechnology applications (health drinks) and organic food (farms, public canteens and restaurants) as well as conventional mass production using industrialised 'productivist' chemical, pesticide, fungicide, herbicide and other conventional control technologies. A once strong but now fading path dependence (historical industry trajectory) was on shipbuilding in Malmö but with the closure of the Kockums yard in the 1980s that led to redundancy and migration of shipyard workers—some to wind-turbine engineering in Jutland, Denmark. By early 2010 the western harbour area had been re-invented as a centre of 'cognitive-cultural' and other service activity in media and multimedia. Activity promoted by the regional development agency also included mobile telephony ('Mobile Heights'), new media ('Media Evolution'), and the Skåne film industry, including computer gaming. An emergent cleantech industry ('Sustainable Hub') and a Systems Resilience initiative ('Training Regions') were also beginning to be visible. The following prioritises regional paradigm resilience while the next regional account, also from Sweden, emphasises regional regime resilience aspects (Cooke 2012).

2.4.2 Mobile Heights

During the 2000s 'Mobile Heights' territory was invaded by rapidly expanding Asian producers from South Korea (Samsung) and China (Huawei). This resilience "shock" (Gunderson and Holling 2002; Folke 2006) led Sony Ericsson to reduce shipments of hardware and re-focus upon managing global services, such as selling network services to mobile telephony suppliers such as Telenord and Telia. To the latter they also sold the extra service of managing the network, clients simply managing billing and cash flow. Accordingly Telia had been cutting employment since the mid-2000s, also no more filing patents. ST Ericsson, the telephony infrastructure arm of the Ericsson Group seemed unlikely to survive as a stand-alone company. Sony Ericsson, the Ericsson mobile telephony joint venture was dissolved. Nokia, Finland's flagship that had a telecoms presence, also nosedived at this time. The main competition for key 'Mobile Heights' member Sony Ericsson was Huawei, which had an office in Lund, Mobile Heights' home base, for the development of basic components for mobile phones. This augments their earlier offices at Kista Science Park in Stockholm, and Gothenburg, employing 250 engineers. Huawei was taking advantage of cutbacks by Ericsson now in Lund that had made hundreds of qualified engineers available. Huawei manufactures across the range from base stations to mobile Internet modems and its own telephone handsets.

Resilience theory, alongside EEG and ECG promises a response, so what was the regional and firm response to these perturbations? Appropriately at regional level an emergent cleantech industry ('Sustainable Hub') and a Systems Resilience initiative ('Training Regions') were beginning to be visible around 2010. Both related to an EU Europe 2020 Grand Challenge shared with Region Västra Götaland to contribute Swedish expertise to building Sustainable Cities (see Fig. 2.1). At firm level, Sony Ericsson rather fruitlessly began evolving "open innovation" relationships with innovative start-ups. Even S.T. Ericsson that was a classic "closed innovation" firm began buying from external suppliers while actively seeking to contract to or acquire them. This remains true in 2015 where Ericsson remains the largest telecoms infrastructure vendor globally, closely pursued by Huawei, meaning Ericsson must

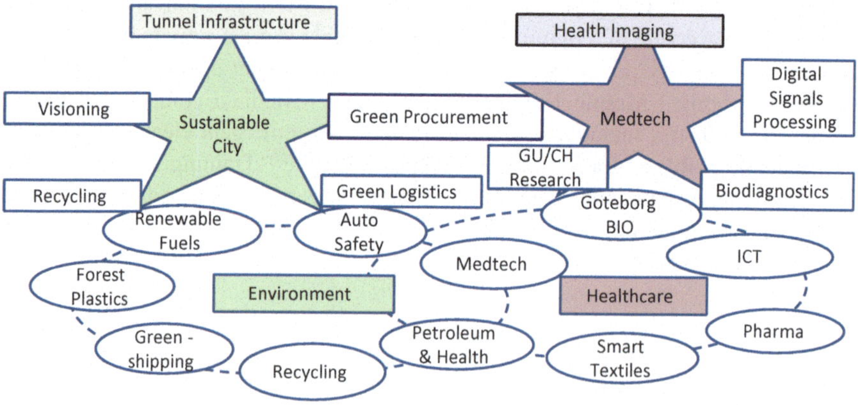

Fig. 2.1 Västra Götaland's 'Iconic Projects' cluster-platform approach. Source: Centre of innovation

seek to emulate Nokia's recovery strategy when it acquired Alcatel-Lucent to consolidate its global infrastructure market. That there were quality entrepreneurial firms in Skåne was testified to by subsequently near-death Canadian mobile telephony firm RIM (BlackBerry) acquiring user-interface maker The Astonishing Tribe (TAT). Moreover, Polar Rose, a Malmö startup which built a facial recognition programme that linked into Facebook photos, was bought by Apple for $29 million, both in late 2010. Other open innovation connections involved 'Mobile Heights' start-ups contracting to AstraZeneca in the Life Sciences platform for remote diagnostics telephony and biosensors. Lateral linkages were also in position with the Media Evolution (Nordic Game) cluster member.

2.4.3 Media Evolution

This Skåne regional cluster concentrated upon 'Convergent Media' otherwise 'New Media'. It promoted the emergence and growth of start-ups in the relevant fields. Most such new firms have entrepreneur leaders with at least 2 to 3 years past experience in larger companies, a minority came from Lund or Malmö University. An example would be Polar Rose which grew out of computer vision research—the analysis of digital images and video—at the Universities of Lund and Malmö. Polar Rose entered the Teknopol Mobile Heights Business Centre in 2004. Teknopol was a tailored business advice agency specialising in start-up activity for the Mobile Heights Business Centre, Sustainable Hub and Life Sciences Business Centre, each of which related to Region Skåne's 'white spaces' cluster-platform programmes. Polar Rose was given an initial loan of €30,000 as a Sony Ericsson spin-out, to develop academically originated face-recognition software. TAT, in 2010 purchased by Research in Motion started in 2002. TAT was to fit its UX-UI, i.e. user experience-user interface 'apps' into BlackBerry's PlayBook and smartphone platform. This was a pioneer user of novel social media forms like 'Crowdsourcing' (Shirky 2010), and 'Crowdfunding' of anything from film projects to start-ups. Accordingly, crowdsourcing was another 'open innovation' response to global, corporate competitive forces impinging upon large Swedish ICT incumbents. A further cross-sector media-ICT innovation link included Qubulus, a system platform for Indoor Positioning on which Location Based Services (LBS) could be developed by Qubulus or by an application developer community through a shared application programming interface (API). The platform aggregates positioning input from proprietary web services and mobile apps to hardware installations. By using the best technology to fit the usage and purpose of the customer case Qubulus can meet user demand and solve the problem of indoor positioning. Crowdsourced positioning activities are a focus in designing space syntax for people flows, shopper movements in retail malls and 'product finder' smartphone applications.

2.4.4 Region Västra Götaland: 'Iconic Projects' Innovation Platform Management

Transversal policies were, at this time, also the characteristic approach taken in Västra Götaland region centred upon Gothenburg. The strategic decision was taken to concentrate initially on meeting the Europe 2020 Grand Challenges of Climate Change and Healthcare, the first not least because the region had been one of the first in the world to publish in 2003 a Climate Change response strategy report 'Gothenburg 2005' involving policies for 'Smart Energy' which then evolved into the strategic Climate Change target of Region Västra Götaland being totally Fossil Fuel Free by 2030. This became known as the 'Gothenburg Model' of the Lisbon Strategy. However, having got the region position on that Grand Challenge worked out in advance gave scope for the new environmental strategy to be down-to-earth and practical. This meant focusing on 'iconic projects' committed to as innovation, learning and collaborative platform management 'laboratories' (Fig. 2.1).

Thus the particularisation of the Climate Change Grand Challenge involved translating it into a 'Sustainable Cities' initiative triggered by an actual large infrastructure commitment to a new tunnel. This brings together numerous regional clusters involved in renewable automotive fuels, forest plastics and petroleum and health. At a more detailed level this assembles pilot projects mixing expertise in cluster firm logistics, public transport, visioning (computer graphics and imaging) and green accounting. It also linked with Chalmers University and specialist firms like Asta AB. A comparable 'Iconic Project' approach was taken in healthcare where the project in question involved a new health complex centred upon a Medical Health Imaging Facility at the University Medical School. This connected transversally to digital signals processing (data compression) and medical diagnostics engineering expertise at Chalmers University and one of its spinout firms Medfield Diagnostics.

2.5 Midi-Pyrénées

The interest here is in an economically strong but over-specialised region that has a narrow path dependence paradigm composed of agro-food, aerospace and healthcare with biotechnology inputs but a strong regional regime that emphasises transversality as a policy model. In the French Pôles de Compétitivité contest it was successful in accessing national cluster-building funding to complement abundant regional and European resources. Remarkably the regional government practises a policy which it calls 'transversalité' in order to populate its narrow regional paradigm with greater path-interdependence. Figure 2.2 represents a process diagram of the regime methodology for inducing transversality from the regional paradigm in a strong way. As explained in Cooke (2012), the steps involved in this prioritised first, the formation of a large and consolidated pool of financial resources derived from the region Midi-Pyrénées itself, the French government and the EU. The next step was

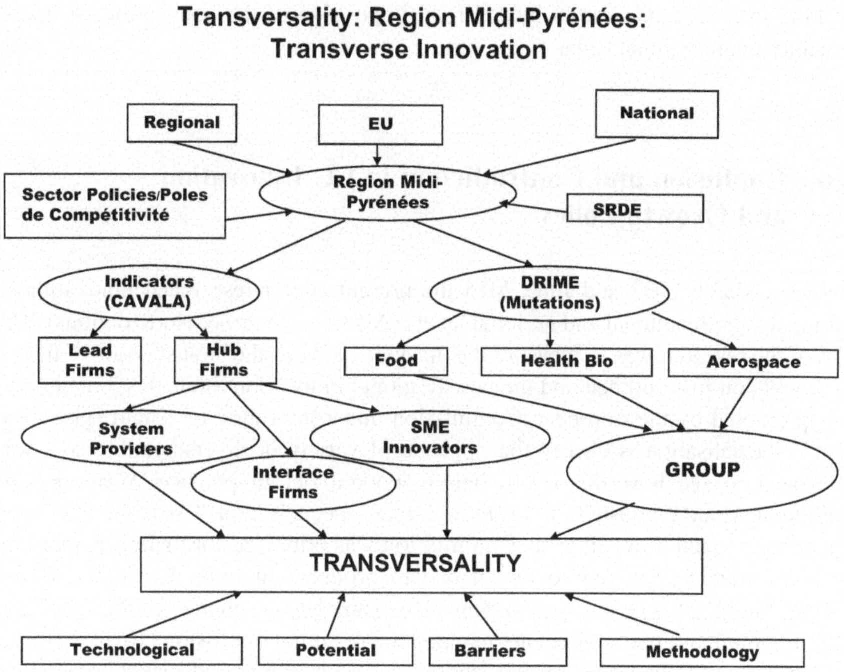

Fig. 2.2 Path inter-dependences and transversality. Source: Own elaboration

to build a methodology for determining how new and greater innovation could be extracted from the region's leading industries through emphasising transversality among them. This led to two parallel exercises, the first (CAVALA) was a statistical review of the strengths and weaknesses of the main clusters and leading firms with respect to innovation and innovation potential. This led to recognition that, in effect, two types of existing and established firm only were likely to be good innovation candidates: lead firms like EADS and Thales in aerospace, and hub firms or 'firmes pivots' i.e. pivotal firms or important systems integrator or aggregator firms in supply chains. To these were added innovative spinout or start-up businesses. Leading candidates from agro-food, aerospace and bio-healthcare were then put in a 'Transversality Group' first to consider methodologies, incentives and rules of the game or 'conventions' by which they might proceed both to talk across sector and cluster boundaries, known to be an especially difficult task where tacit knowledge is concerned (Janowicz-Panjaitan and Noorderhaven 2009). In these group discussions the key focus was on technology, its known properties and cross-pollination potentialities, barriers to innovation from cognitive, research or resources viewpoints and, as noted, methodologies by which firms might find each other despite their apparent unrelatedness in order to generate regional innovation through exploiting relatedness. This is a new, rather typically French top-down model of seeking to induce innovation by a formal imposition of the conventions of transversality upon regional firms.

2.6 Confusion and Contradiction in EU Innovation and Growth Policy

Between March 2013 and June 2015 the present author researched innovation in Portugal at both national and regional levels (Algarve, Centro & Norte regions). The aim of the research was to measure the distance between the "transversality" theory of innovation just outlined and the new Regional Innovation Strategies 3 methodology promoted by the European Commission under the rubric of "smart specialisation". Specialisation is clearly the opposite of variety or diversification, so I was interested to see how this contradiction worked out in practice. Were regions sacrificing valued industries to promote "smart specialisation"? Was the idea even understood? And how, after the Commission was criticised for its linear, sectoral, specialisationist approach so that it had to propose (in footnotes) that "related variety" and DUI-type innovation were also examples of "smart specialisation" did their regional and national clients manage the resulting confusion (Kroll 2015)?

This proved to be an interesting laboratory for observing multi-level governance (MLG) tensions—from region to nation to supra-nation (EU) levels of interaction. The context is unique in that a slow-moving, cumbersome and—as many see it—spatially myopic and conceptually chaotic European Commission belatedly sought to induce a new, post-program budgeting and linear regional economic development model to promote growth while imposing major constraints in the form of austerity policy, budget cuts and draconian debt repayment conditions. At its worst, the austerity strategy had massively impoverished Eurozone member Greece and while Portugal emerged from the imposed fiscal straitjacket without the devastating results experienced there, the hallmarks of contradictory thinking about how the EU "believes" it promotes growth by imposing conditions that ensure its opposite remain evident.

In brief, the studied regions and even, to some extent the state, ignored the precepts of specialisation and pursued the common-sense potential of optimising their regional diversity to promote regional innovation. This meant Algarve aiming not only to escape its narrow over-specialisation in "sun & beach" tourism by pushing for DUI applications of renewable energy, marine biology, ICT and creative industries to diversify their tourism and—with the help of a regional innovation agency—developing new industries, some embodying STI-type innovation from universities and research centres. These could be outside Algarve if necessary. However, it was a very "horizontal" set of aspirations. Centro and Norte already had high "related variety" scores as judged by the Portuguese National Research Council (FCT 2013) so they used matrix methods to identify crossover innovation opportunities and projects in biotechnology, flexible manufacturing systems, robotics, renewables and footwear amongst other intersecting "innovation platforms". In the last two cases their strategies were accepted by the state but it retained control of project evaluation (dependent on the EU Regional Operational Programmes into which RIS3 allocations fit). But for Algarve the state's innovation ministries and agencies (also the other regions) opposed their diversity plans on grounds of "lack of critical mass" thus condemning Algarve to remain specialised but not especially smartly so. A better governance model for regional innovation was, however, approved but not a full-blooded regional innovation agency.

So the adoption of a "specialisationist" model in the field of ERDF allocations via ROPs to subsidise regional innovation and growth was rejected by Portugal's regions and even in limited ways by the state. In its stead, diverse regions either sought to initiate, or where conditions were more evolved, consolidate growth opportunities and gains by adopting regional diversity through building on the concept of related variety and the fashioning of "transversal" innovation policy (Cooke 2016). That this was given approval in the RIS3 documentation promoting "smart specialisation" merely underlies the conceptual confusion and spatial myopia of the EU and its Commission already mentioned. This shows the EU and even its member-states are slow-moving and backward thinking policy action entities. Even weak regional administrations such as those anatomised above can respond and in limited ways even anticipate needed economic policy actions more swiftly. However, "at the edge of chaos" as understood in EEG and complexity theory, where change is imminent or unavoidable "fortune favours the prepared mind" as Louis Pasteur saw it. Centro and Norte saw clear advantages in exploiting innovation opportunities arising from past R&D infrastructural investments and their "sensemaking" crossover thinking was hard to oppose by the state. Algarve had great difficulty extracting its future innovation profile from the specialised "sun and beach" "frame" endowed upon it by its state and fellow regions. The key problem lies in institutional failure by big, slow organisations like the EU and member states to leave their neoclassical industrial economics comfort zone and embrace the full meaning of innovation, which is that it is recombinant, interactive and unconfined to a sector or even a cluster. Rather it is geographical, interactive and based on "crossover" innovation at interfaces.

2.7 Conclusions

It is clear that the ECG "Transversality" perspective can be considered successful at pathbreaking in three significant dimensions. First, the theoretical sophistication of its approach evolves its evolutionary economic geography (EEG) origin in an approach to a primary position from the viewpoint of advanced regional analysis. This utilises evolutionary concepts from economic geography, complexity and resilience theory such as the multi-level perspective, complex adaptive systems, external shocks and internal perturbances, preadaptation, adjacency, cognitive reversal, relatedness, proximity, path dependence and transversality in a coherent, innovative and intellectually penetrative way. Much further research is likely to follow into the explanatory validity of this non-reductionist, non-predictive evolutionary framework. Kauffman (2008) presents this perspective as "lawless" in the sense that it is beyond the paradigm exemplar of neoclassicism in economics, which derives mechanistically from physics. Since life forms cannot be predicted this approach escapes the strictures of that reductionist frame. The second major contribution of the findings on knowledge flows and innovation for the future concern its critical reflections on numerous inadequately scrutinised aspects of innovation theory. Accordingly, innovation is now better-specified as the key element of any evolutionary, especially ECG, growth model. Finally, the theoretical and empirical results have shown how relatedness and transversality are practised in the actualité and may be empirically observed by firms and policy agencies seeking or charged with enhancing business and regional innovation. This strongly suggests the validity of Kurt Lewin's observation that 'there is nothing so practical as a good theory'.

Acknowledgements The following colleagues assisted me with background information and advice in composing this chapter. In Sweden, Lennart Svenssen, Arne Eriksson, Carin Dahl, Helina Nilsson. In Italy, Marco Marchese, Jonathan Potter. In Portugal, Vitor Corrado Simoes, Artur da Rosa Pires, Ana Rita Cruz, Hugo Pinto Antonio Ramos Teresa Jorge, Alexandra Rodrigues, Jorge Brand*ão*, Alexandre Almeida and Rui Montero. All are thanked and the usual disclaimer applies.

References

Arthur, B. (1994). *Increasing returns and path dependence in the economy*. Ann Arbor: University of Michigan Press.

Arthur, B. (2009). *The nature of technology*. London: Penguin.

Balconi, M., Brusoni, S., & Orsenigo, L. (2010). In defence of the linear model: An essay. *Research Policy, 39*, 1–13.

Chapain, C., Cooke, P., De Propris, L., MacNeill, S., & Mateos-Garcia, J. (2010). *Creative clusters and innovation: Putting creativity on the map*. London: NESTA.

Cooke, P. (2012). Relatedness, transversality and public policy in innovative regions. *European Planning Studies, 20*(11), 1889–1907.

Cooke, P. (Ed.). (2013). *Reframing regional development*. London: Routledge.

Cooke, P. (2016). Four minutes to four years: The advantage of recombinant over specialised innovation – RIS3 versus "smartspec". *European Planning Studies, 24*(8), 1494–1510.

Cooke, P., & Schwartz, D. (2008). Regional knowledge economies: An UK-EU and Israel perspective. *Tijdschrift Voor Economische en Sociale Geografie, 99*, 178–192.

Cooke, P., De Laurentis, C., MacNeill, S., & Collinge, C. (2010). *Platforms of innovation*. Cheltenham: Edward Elgar.

David, P. (1985). Clio and the economics of QWERTY. *American Economic Review, 75*, 332–337.

Dew, N., Sarasvathy, S., & Venkataranam, S. (2004). The economic implications of exaptation. *Journal of Evolutionary Economics, 14*(1), 69–82.

FCT. (2013). *An analysis of the Portuguese research and innovation system: Challenges, strengths and weaknesses towards 2020*. Lisbon: Fundação para a Ciência e a Tecnologia/National Research Council.

Felsenstein, D. (2011). Human capital and labour mobility determinants of regional innovation. In P. Cooke, B. Asheim, R. Boschma, R. Martin, D. Schwartz, & F. Tödtling (Eds.), *The handbook of regional innovation & growth* (pp. 119–131). Cheltenham: Edward Elgar.
Folke, C. (2006). Resilience: The emergence of a perspective for social-ecological systems analysis. *Global Environmental Change, 16*, 253–267.
Frenken, K. (2006). *Innovation, evolution & complexity theory*. Cheltenham: Edward Elgar.
Garud, R., & Karnøe, P. (2001). Path creation as a process of mindful deviation. In R. Garud & P. Karnøe (Eds.), *Path dependence and creation* (pp. 1–38). London: Lawrence Erlbaum.
Geels, F. (2007). Analysing the breakthrough of rock 'n' roll (1930–1970): Multi-regime interaction and reconfiguration in the multi-level perspective. *Technological Forecasting & Social Change, 74*, 1411–1431.
Grabher, G. (Ed.). (1993). *The embedded firm: On the socioeconomics of industrial networks*. London: Routledge.
Gunderson, L., & Holling, C. (Eds.). (2002). *Panarchy: Understanding transformations in human and natural systems*. Washington, DC: Island Press.
Hughes, T. (1977). Edison's method. In W. Pickett (Ed.), *Technology at the turning point* (pp. 5–22). San Francisco: San Francisco Press Inc.
Jacobs, J. (1961). *The death & life of great American cities*. New York: Vintage.
Janowicz-Panjaitan, M., & Noorderhaven, N. (2009). Trust, calculation and interorganizational learning of tacit knowledge: An organizational roles perspective. *Organization Studies, 30*, 1021–1044.
Jensen, M., Johnson, B., Lorenz, E., & Lundvall, B. (2007). Forms of knowledge and modes of innovation. *Research Policy, 36*, 680–693.
Johnson, M. (2010a). *Seizing the white space*. Boston: Harvard Business Press.
Johnson, S. (2010b). *Where good ideas come from*. New York: Riverhead.
Kauffman, S. (2008). *Reinventing the sacred*. New York: Basic Books.
Kline, S., & Rosenberg, N. (1986). An overview of innovation. In R. Landau & N. Rosenberg (Eds.), *The positive sum strategy* (pp. 275–305). Washington, DC: National Academy Press.
Kroll, H. (2015). Efforts to implement smart specialization in practice – Leading unlike horses to the water. *European Planning Studies*. https://doi.org/10.1080/09654313.2014.1003036
Krugman, P. (1991). *Cities in space: Three simple models* (NBER working paper 3607). Cambridge: National Bureau of Economic Research.
Krugman, P. (1995). *Development, geography & economic theory*. Cambridge: MIT Press.
March, J. (1991). Exploration and exploitation in organisational learning. *Organization Science, 2*, 71–87.
Martin, R. (2010). The Roepke lecture in economic geography – Rethinking regional path dependence: Beyond lock-in to evolution. *Economic Geography, 86*, 1–27.
Martin, R., & Sunley, P. (2006). Path dependence and regional economic evolution. *Journal of Economic Geography, 6*, 395–438.
Martin, R., & Sunley, P. (2010). The place of path dependence in an evolutionary perspective on the economic landscape. In R. Boschma & R. L. Martin (Eds.), *Handbook of evolutionary economic geography* (pp. 62–92). Cheltenham: Edward Elgar.
Myrdal, G. (1957). *Economic theory & underdeveloped regions*. London: Duckworth.
Nunes, S., & Lopes, R. (2015). Firm performance, innovation modes and territorial embeddedness. *European Planning Studies, 23*(9). https://doi.org/10.1080/09654313.2015.1021666
Nystrom, G., Razaq, A., Strømme, M., Nyholm, L., & Mihranya, A. (2009). Ultrafast, all-polymer, paper-based batteries. *Nano Letters, 9*, 3635–3639.
Page, S. (2007). *The difference*. Princeton: Princeton University Press.
Pisano, G., & Verganti, R. (2008). Which kind of collaboration is right for you? *Harvard Business Review, 86*, 78–86.
Rosenstein-Rodan, P. (1943). Problems of industrialisation of eastern and south-eastern Europe. *Economic Journal, 53*, 202–211.
Schumpeter, J. (1934). *The theory of economic development*. Cambridge: Harvard University Press.

Scott, A. J. (2008). *Social economy of the metropolis: Cognitive-cultural capitalism and the global resurgence of cities*. Oxford: Oxford University Press.

Shirky, C. (2010). *Here comes everybody*. London: Penguin.

Strambach, S. (2010). Knowledge-intensive business services. In P. Cooke et al. (Eds.), *Platforms of innovation*. Cheltenham: Edward Elgar.

Veblen, T. (1898). Why is economics not an evolutionary science. *Quarterly Journal of Economics, 12*, 373–397.

Villani, M., & Ansaloni, L. (2010). A theory-based dynamical model of exaptive innovation processes. In P. Ahrweiler (Ed.), *Innovation in complex social systems*. London: Routledge.

Vrba, E., & Gould, S. (1982). Exaptation – A missing term in the science of form. *Paleobiology, 8*, 4–15.

Chapter 3
Evolutionary Resilience Shifting Territorial Development Paradigms

Carlos Gonçalves

3.1 Introduction

Combining two interpretations, resilience as a normative to achieve sustainability (Pinho et al. 2008) and resilience as a dynamic aim (as opposed to sustainability which appears to be based on some inflexibility) we will consider to what extent the designation "sustainability" is reinforced when seen from the paradigm of evolutionary resilience.

To build resilience into systems that support an urbanized world requires planning with a wide scale of uncertainty and that communities be enabled to interpret movement(s) of the wave, rather than to be carried away by the swirl that constitutes in its interior (Gross 2008). Planning evolutionary resilience gains implies that the process of transition should be analysed and the mechanisms of paradigm change should be assessed according to a new vision of sustainable development "where resilience implies normalizing environmentally sustainable practices. This is more closely aligned with the original drive towards sustainability seen in the Brundtland Report (WCED 1987), and thus we return to seeking a more profound understanding of how sustainability will work in practice in the complexity of our cities" (Collier et al. 2013, p. 6).

Following this reasoning, in the next sections, we will develop defining elements of the latitudes where the paradigm of evolutionary resilience widens, opening perspectives which direct the reader to its use in the process of territorial development.

Hence, we circumscribe the scope of this chapter as follows: in the first section the etymological origins of the concept are approached, in the second we point out elements that allow us to understand how the "science of resilience" was expanded.

C. Gonçalves (✉)
Institute of Geography and Spatial Planning. Universidade de Lisboa, Lisboa, Portugal
e-mail: c.goncalves@campus.ul.pt

H. Pinto et al. (eds.), *Resilience and Regional Dynamics*, Advances in Spatial Science, https://doi.org/10.1007/978-3-319-95135-5_3

Then, and in order to question its so-called consistency, the validity of this perspective on the processes of development is discussed: is it a coherent conceptual framework? Is it a new paradigm able to strengthen sustainability as a civilizational goal or is it just an illusion? Joining the three levels (etymological origins, scientific rooting process, and epistemological construction), we will try to add to the debate about the contribution that this approach can give to answer the need of repositioning development models.

3.2 Etymological Origins

The etymological foundations of the term 'resilience' and its variations open a wide and diverse window onto where its origin can be seen and its different meanings, including those of most recent research. Looking at the historical uses to which the word was applied to was proposed by Alexander (2013), and originally it was meant to highlight that the concept was conceived by Stanley Crawford Holling in an article in 1973 (Holling 1973). In many cases, literature grants the creation, or at least, the establishment of the resilience concept to this author and his work entitled "Resilience and Stability of Ecological Systems" (Holling 1973). Berkes (2007), Djalate (2011) Goldstein and Brooks (2006), whose works are mentioned by Alexander (2013), are just some of many authors (almost all) who address this idea.

However, terms like *"resilire"* or *"resilio"* (resilience) indicate its origin from the Latin, associated with the meanings: return to a starting point, jump back, recover, retreat (*"bounce"* or *"bouncing back"*). This is one of the elements that sustain the existence of a significantly more extensive base. In classical literature lies another one[1]: Seneca[2] uses the word as something that has the ability to jump or leap. In *Metamorphoses*, Ovideo associates it with something that shrinks, wanes or contracts (*"shrink"*). Quintillian[3] applies the idea that underlies it to categorize something intended to be achieved.

Up until now, the term splits into three meanings: something that takes up a previous position; something that jumps or leaps (frogs, for example); and an entity whose attribute is to resist (the latter appears in Cicero, in his "*Orations*") (Alexander 2013). These implications have been maintained much through the resonance that the proverbs of *St. Jerome* gained (347–420 AD). However, in many cases, the connotations are negative. For example, the meaning of return corresponded to situations of failure, of sadness, of loss. The same happens when someone refuses reality, preferring to resist, retreat, not accept or alienate oneself from the world around him/her (Alexander 2013).

[1]At this time the term appears in texts of Seneca the Elder (54 AC-39 DC), Pliny the Elder (DC23-79), Ovideo (43 AC-18 DC), Cicero (AC106-43) and Livy (59 AC-17 DC).

[2]Adolf Gottlieb Kiessling, aedibus B. G. Teubneri.

[3]Marcus Fabius Quintilianus, Istitutio Oratorio, 12, 10.56.

Later, the word appears in medieval French. This time, the verb *"résiler"* carried the meanings: to retract, retreat (*"to retract"*) and to cancel, to give up (*"to cancel"*). From here the bridge to the English use (*"resile"*) is made, which first appears in the state papers of Henry VIII (in 1529) where it is used to describe situations of retreat, resuming former positions, or abandonment. The term was used at the time of the conflicts that this monarch had with Queen Catherine of Aragon (1485–1536).

According to Alexander (2013), the first use of the word resilience ("resilience") in a scientific context was made by Sir Francis Bacon in 1625 in the publication *Sylva Sylvarum* (Bacon 1625, p. 66). To describe the types of echoes and the effects they produce, the author used the term "resilience" (Fig. 3.1). The definition, recorded in a dictionary, comes up in 1661 by the hand of Thomas Blount. Blount gathered 11,000 definitions to terms he considered far-removed from daily language (i.e., "*hard words*") in a publication entitled "*Glossographia*".

Despite some variations, such as those that German assigned to it (*Elastizität*), in the first half of the nineteenth century, the term resilience was still clung to the notions of to recover, to resume, to restore, but also began to be connected to concepts such as elasticity, instability, and volatility. Its usage to describe the ability to withstand adversity through "mental strength" appears in 1839 (Bell 1839, p. 344, quoted by Alexander 2013). The context that marks its use when describing the ability to withstand the impact of adversity is noticed in descriptions of the resourcefulness and tenacity that the Japanese expressed to resist the two earthquakes that hit the city of Shimoda (located southwest of Tokyo) in 1854 (Tomes 1857, p. 379 cited in Alexander 2013).

In Fig. 3.2 the chronological path the concept undertook in different areas of knowledge is represented.

Its application in the field of engineering materials began in 1858 when William J. M. Rankine (1867) employed the term to describe the ductility of steel beams. From this moment, its adoption spread into different areas such as: coronary surgery,

245 "The *Eccho* commeth as the *Originall Sound doth*, in Round Orbe of
Aire: It were good to try the Creating of the *Eccho*, where the Body
Repercussing maketh an Angle : As against the Returne of a Wall,
&c. Also we see that in *Mirrours*, there is the like Angle of Incidence,
from the Object to the Glasse , and from the Glasse to the Eye. And if you
strike a *Ball* side long, not full upon the Surface, the Rebound will be as
much the contrary way; Whether there bee any such *Resilience* in
Eccho's, (that is, whether a Man shall heare better , if he stand aside the
Body Repercussing, than if he stand where he speaketh, or any where in a
right Line betweene;) may be tried. Triall likewise would be made,
by Standing nearer the Place of Repercussing , than he that speaketh ;
And againe by Standing further off , than hee that speaketh ; And so
knowledge would be taken, whether *Eccho's* , as well as *Originall Sounds,*
be not strongest neare hand".

Fig. 3.1 Use of the word resilience in the book *Sylva Sylvarum Or The Naturall Historie* of Francis Bacon. Source: *Sylva Sylvarum Or The Naturall Historie*, Francis Bacon (1625, p. 66)

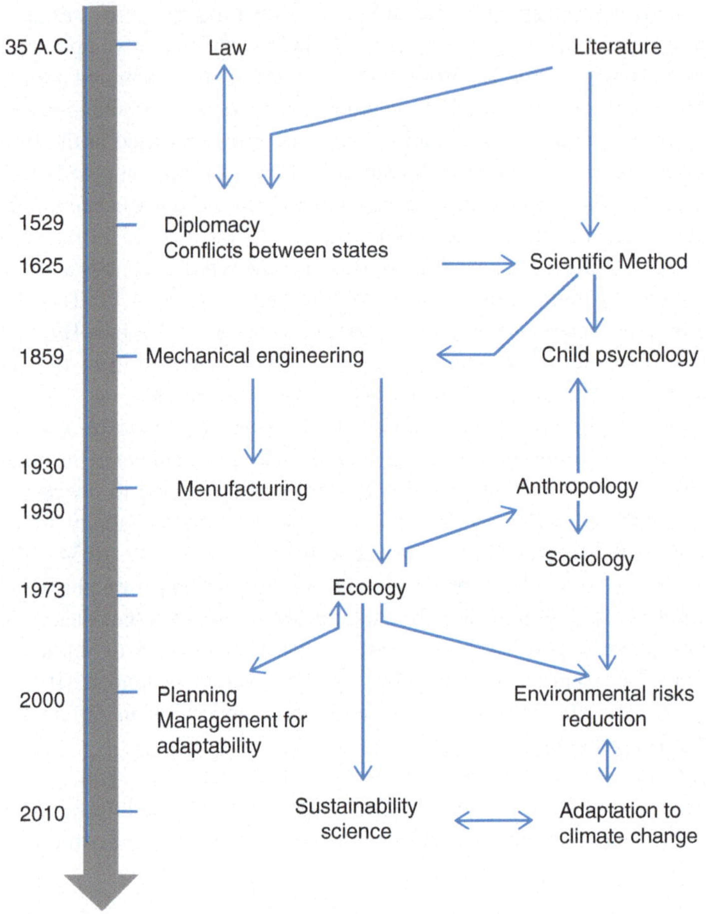

Fig. 3.2 Schematic representation of the evolution of the use of the term resilience. Source: Based on Alexander (2013)

anatomy or horology, or even to qualify certain threads (and techniques) used in weaving. In this context, resilience carries on a dialogue with ductility and stiffness/strength to describe the properties of some metals/materials. This is where the analogy was applied to the behavior communities' displayed when suffering under extreme events. Robustness is the ability to withstand the shock while ductility is associated with the power to absorb impact.

In the 50s, its use by psychology began, which amplified its popularity in the 80s, with particular emphasis on the monitoring of diseases that affect children (Goldstein and Brooks 2006). In the 90s, it was gradually transferred from ecology to human ecology and, henceforth took its place in social sciences. In what concerns this last "leap", it occurred via the input of economists such as Batabyal (1998), and geographers like Adger (2000). The last 40 years were the stage for this development and consequent amplification of the number of concepts that depart from or flow into the term resilience as a paradigm.

Table 3.1 presents some elements that comprise a synthesis of the different settings of the concept. Its origin is pointed out as well as a summary of the principles underlying them, the focus of each and in the last two columns, the way they respond, according to the nature of the shock suffered. For this purpose, we use a region's hypothetical behavior in employment. This table will be used as a guiding principle for the following points in which we will be developed in more detail each of the configurations of the concept and will also discuss how the dialogues and transitions from one configuration to the other occur (reconfigurations).

The first definition of resilience is developed around three dimensions: resistance, stability, and equilibrium. Physical resilience (within the sense connected to engineering) refers to stability and assumes that it is possible to determine the position where it can be reached. To the resistance ability that a given structure or material needs so that does not drift apart from a position of lesser effort, it is added the speed (in cases where dislocation was not avoidable) with which the point of origin is retrieved (Pendall et al. 2010, p. 72 citing Berkes and Folke 1998).

Imported from ecology and widely used in medicine (especially the approach to the regeneration process after exposure to traumatic experiences), the concept of resilience is defined as "the ability to return to the shape or original position after being bent, compressed or stretched out. It refers to the elasticity and ability to quickly recover from illness, depression, anxiety, or something similar" (Schroll et al. 2009, p. 1). However, besides these areas, its use in the study of ecosystems as well as the multiple relationships involved in the research done within the field of social sciences are some further ways (among others) through which this type of approach has lately been followed.

When drawing up the itinerary to be followed in this study and in order to overcome all the nuances of the conceptual resilience aggregate, if we take note of the proposals gathered in Table 3.2, there are at least three directions. In one, descriptive approaches to the concept are found, distinguishing the plans that result from social sciences and those which derive from natural sciences; in another, we can glimpse the arrangements done in order to grant inseparability to the relationship between natural and social systems; finally, needs emerge to open normative possibilities for the sustainability of development processes.

With the theoretical development and the widespread effort in practical use, the concept takes shape and complexity, serving new purposes. To put it plainly, and corroborating Martin's proposal (2012) as well as the additional contributions observed by Pendall et al. (2010), we can say there are three fields in which the resilience concept was defined (and redefined): (i) as property developed by engineering; (ii) as a trivial characteristic of ecosystems mechanisms; (iii) as an adaptive property associated with socio-economic and socio-environmental systems where complexity is central, namely the one which is introduced by the human factor.

In the next section we tried to set elements to answer the questions: is the approach to territorial development, which makes use of resilience, working with a new conceptual framework orientated towards a new paradigm, or is it merely a new

Table 3.1 Introductory elements to understand the concept of the three configurations of resilience: engineering, ecologic, adaptive

Configuration of the concept	Scientific area of origin	Principles	Focus	Nature of disorder	
				Slow progression	Sudden Impact
Resilience of structures and materials	Physics	The system returns to the starting point reassuming the stage of equilibrium, or configuration, after being subjected to a shock It is based on the resistance to shock and in the stability in reference to an equilibrium The system resists changes to its "normalcy" and when these happen, it has the ability to swap back to it Measures: recovery time	Unique equilibrium	Maintain the established norms, understood as "natural" Example: Full employment, despite the present weakness in the economic base	Resumes normalcy Example: The employment level is reset on the path that it followed prior to the disturbance
Ecologic resilience	Biology; Ecology	The magnitude of the shock or disorder that the system can absorb, prior to being destabilized and propelled to another stage of stability. It is based on systemic analysis and considers the *feedback* generated in the system, which is seen as a whole Measures: robustness disturbance absorption capacity	Multiple equilibrium	Improves performance Example: Employment levels exceed those that existed before the shock	Establishes a "new normality" Example: employment growth positions itself in different trajectories benefiting from conditions arising from the shock
Adaptive resilience	Complexity theory; Adaptive systems	Capacity (anticipated or reactive) of reorganizing shape and/or function in order to minimize the destabilizing effect of a shock Measure: Vulnerability to surprise; stress and shock; flexibility of responses	Adaptive complex systems	Incremental adaptation (or continued) Example: The adjustment reduces the loss of employment and its volatility	Incremental adaptation (or continued) Example: the (re)distribution of resources, power, and opportunities support levels of acceptable employment trajectories

Source: Adapted from Pendall et al. (2010, p. 81) and Martin (2012)

Table 3.2 Resilience settings

Categories and classes			Definitions
Descriptive approaches	Ecology	Original formulation	"Measure of the persistence of the systems and their ability to absorb change and disturbance maintaining the same relationship between populations (or between state variables)" (Holling 1973, p. 14)
		Extensions to the original formulation	"The magnitude of disturbance that can be absorbed before the system changes its structure by changing the variables and processes that control their behavior" (Gunderson and Holling 2002, p. 4) "The ability of a system experiencing shocks essentially maintaining the same function, structure, reactions, and therefore preserving the identity" (Walker and Salt 2006, p. 2)
		Three capacities	"Capacities: (i) to absorb disturbances; (ii) to generate self-organization, and (iii) to trigger learning and adaptation" (Walker et al. 2002)
		Fourth domains	"(1) latitude (width of the domain), (2) resistance (height of the domain), (3) the precariousness, (4) relationship between scales" (Folke et al. 2004, p. 573)
		Heuristic system	"Quantitative property that in ecosystems, change in a dynamic way, occurring on every hierarchy level" (Holling 2001)
		Operational approach	"Resilience of what and for what?" (Carpenter et al. 2001)
			"The ability that the system demonstrates to maintain its identity in frames of disturbances, change and internal and external shocks" (Cumming et al. 2005)
	Social sciences	Sociology	"The ability that groups or communities manifest to deal with tensions and external disturbances resulting from social, political and environmental changes" (Adger 2000, p. 347)
		Socioeconomic	"Transition probability between the states as the function of the balance between production activities and consumption patterns, conditioned by the action of decision-makers" (Brock et al. 2002, p. 273)
			"The capacity present in the system to withstand shocks in the functioning of the market or the environmental structure without compromising the ability to distribute resources in an efficient way" (Perrings 2006, p. 418)
Hybrid approaches		Ecosystem service providers	"The intrinsic ability that the ecosystem has to maintain the desired environmental services, even in unstable environmental situations induced by human activities" (Folke et al. 2002, p. 14)
		Social-ecological systems	"The ability that a social-ecological system has to absorb recurrent disorders (...) maintaining the essential structures, processes, and feedback" (Adger et al. 2005, p. 1036)
		Resilience as a reference	"Perspective or approach, able to analyse social-ecological systems" (Folke 2006)

(continued)

Table 3.2 (continued)

Categories and classes			Definitions
Normative approaches		Metaphorical sense	"Flexibility designed in the long term" (Pickett et al. 2004, p. 381)
		Reinforced sustainability	"Maintenance ability in the long term of a given natural capital" (Ott and Döring 2004 p. 213f)

Source: Adapted from Brand and Jax (2007)

illusory path? Furthermore, how has, what began to be called "science of resilience", been expanded? Let us start with the latter of the two so that, through it, and immediately afterward, we can work on the answer to the first question.

3.3 The Widening of "Science of Resilience"

3.3.1 Configuration of a Research Field

The perspective that resilience is a consistent resource for understanding the dynamics of socio-environmental systems is gaining strength (Thapa et al. 2010; Anderies et al. 2006). Growing interest and undefined contours are two repeatedly summoned ideas for the first paragraphs of the literature devoted to resilience. The growing interest is said to result from its plasticity, the spreading of adopted meanings, and perhaps above all the rising advantages of crisis (in most cases apparently unexpected) as a structural element of contemporaneity. Its utility is important as a structure of thought and intervention. It is also relevant as a rationale that allows both questioning the unleashed or accelerated crises episodes and explaining why a paradigm statute guider of governance models is assigned to these crises periods.

Using the *Social Science Citation Index*, it is clear the evolution of the number of citations of the word 'resilience' occurred only in the decade measured between 1997–2007. The exercise's result is transposed to an increased measure of about 400%, taking into account the reported timeframe (Swanstrom 2008).

Using another type of methodology, Xu and Marinova (2013) provided more detailed information, covering the timeline between 1973 and 2011. The authors built a quantitative and qualitative analysis based on the citations generated in the publications devoted to the subject and offered an insight into the extent of what is produced and the community amplitude dedicated to this type of studies. The information is generated using the following databases: "*Google Scholar*", "*Web of Science*" and "*Scopus*", which, in addition to '*resilience*', retrieved information to eleven combinations of keywords.[4] Since the results are delimited within the area,

[4]The combinations were as follows: *"ecological resilience", "economic resilience", "social resilience", "resilience and sustainability", "resilience and sustainable development", "resilience and social-ecological systems", "social-ecological resilience", "resilience and environment",*

they provide an image of the more restricted core of the 'community' that has been dedicated to uniting this branch of science.

The foundation of the "*Resilience Alliance Network*" (in 1999) and of the Journal "*Ecology and Society*" (in 2006) indicate two moments of increase in both the number of publications and in the created citations. Other landmarks such as the "*Millennium Ecosystem Assessment Reports*" (published in 2005), the "*Stern Review*" (in 2006), and the "*Intergovernmental Panel on Climate Change (IPCC)'s 4th Assessment Report*" (in 2007) nourished the interest in the area.

Despite the importance of the extension of studies dedicated to ecological systems, the increase of publications (especially since 1995) has become more transversal. However, in most cases, the organizational matrix continues to refer to ecology, as shown by the fact that most publications have their origin in the journal "*Ecology and Society*".

More recently greater enthusiasm is perceived in areas to which economy[5] and sustainability conjoin. To understand and assess sustainability from the perspective of complex systems and given the unpredictability of the global transformations that are taking place, we are obliged to have a grasp on the research that this area of study has produced. Although this path is already signaled, the route taken by way of social-ecological, socio-economic, and urban systems is still short (Xu and Marinova 2013).

From the geographic distribution analysis (by country) of the production of cited articles, the results of the USA, Australia, and Sweden are highlighted. Most of the case studies are also located in the USA (25.4%) while Europe is the second continent regarding the concentration of case studies (21.8%).

The weight measurement of this type of research results from the balance between the number of researchers who are dedicated to this field of studies compared to the total[6] each country has. However, we removed the countries where the total number of researchers is reduced, which might lead to erroneous results since the amount of those who engage in the area results in oversize. Examples of this are Australia (2,6 per each 1000) and Sweden (1,8 per each 1000) whose intensity is more noticeable. In most cases, though, the most substantial part of the production still develops in the ecology area and in the theoretical discussion.

In Portugal, there has been an incipient participation with the results pointing to 9 researchers in this area, which represents 0.024 per each 1000. In most cases, these scholars were devoted to the theoretical discussion (4), only 3 are dedicated to

---

"resilience and natural resources" and "resilience and assessment". The criteria for the respective publications to be included in the study were: (i) that these expressions be part of the title; (ii) that they are included in the list of keywords; or else, (iii) that they appeared at least three times in the abstract. This restraint in the criteria leaves out all publications that present synonyms or antonyms of the term 'resilience', excluding also those that did adopt the English language.

[5]Noting this interest, the Cambridge Journal of Regions, Economy and Society, dedicated a volume (3) in March 2010 to the theme: "The Resilient Region".

[6]This exercise is done using the UNESCO's database (http://www.uis.unesco.org/ Pages/default.aspx?SPSLanguage = EN)", (Xu and Marinova 2013).

approaches close to the economy and 2 develop work on a social level. There is no record at this stage of researchers using transversal approaches like the one we have tried to propose in our studies.

Despite being the country with more productivity, in the USA the *University of California* represents only 10% of the total research done within this paradigm, meaning that in this nation the research is not concentrated. On the contrary, it is spread over several institutions. The same is true in countries such as the United Kingdom, Germany, France or Canada. On the extreme opposites are the cases of India, Sweden, and Norway, where the *Institute for Social and Economic Change*, the *Stockholm University* and the *University of Oslo* gather 64, 59 and 54% of what is produced in these countries (Xu and Marinova 2013).

Taking into account the increase of instability(ies), Xu and Marinova (2013) call upon what is said by Walker et al. (2004), Adger et al. (2005) and Folke (2006) to strengthen the idea that the resilience paradigm is important (perhaps of the greatest importance) to interpret and intervene in the relationship between environmental and human disturbances in order to strengthen sustainability, which is regarded as a civilizational aim.

In a more circumscribed approach, concerning the growing interest in further developing the resilience of urban systems, the value of publications focused on this particular segment are equally symptomatic of the remarkable leap that has occurred in the number of publications, particularly since 2001. In that year the number of publications in this area was around 20 papers. In 2011 the value increased to 240.

The theme has gradually occupied center stage in the discussion of urban policies. Some recent examples are: the AESOP (*Association of European Schools of Planning*), which in 2010 dedicated the first symposium to the theme "*resilient cities*" in Stockholm (date after which a thematic group was created); in 2013 in Dublin the *Joint AESOP/ACSP Congress*, under the title "*Planning for resilient cities and regions*" took place; the *Regional Studies Association Global Conference 2014* took place in Fortaleza (Brazil) discussing the theme: "*From vulnerability places to resilient territories: the path to sustainable development*"; and the *Regional Studies Association European Conference* held in June 2014 in Izmir (Turkey) devoted to "*Diverse regions: building resilient communities and territories*".

3.3.2 Widening the Spectrum of Practical Application

Besides the academic interest, it is also noticeable there is a growing assimilation of the resilience paradigm by technical communities, political structures and non-governmental organizations that, whether in one or another, direct their performances to various scales (Béné et al. 2012). Examples of this are the initiatives of: the *World Bank Social Protection* that in the "*Labour 2012–2022 Global Strategy*" defines as a goal increasing resilience, equity and opportunities for countries with low and average salary levels; the *United Nations Office for Disaster Risk Reduction* (UNISDR 2012), which launched the initiative "*Making cities Resilient*"; the *World*

Food Programme and the *Swiss Agency for Development and Cooperation* precursor of the "*Resilience Project*"; the *US Agency for International Development*, which has an on-going initiative called "*Resilience Week*"; and the *Australian Aid Agency* that is implementing the environmental strategy "*Building Resilience, Sustaining Growth*" (Béné et al. 2012, p. 8). The *European Spatial Planning Observatory Network* has an on-going project—the *ECR2* project—Economic Crisis: Resilience of Regions[7]—devoted to the assessment of the impacts asymmetry of the current crisis process that started in 2008 in regions of the European Union. The aim of this project is to explain why some regions within the EU suffered impacts of small significance while others have recovered the development paths, whereas in other cases the regression processes are increasing at a deeper and deeper rate and have extended in time. This team's goal is to propose guidelines for economic structures within the European Union, its countries, its regions and its cities, so that they can increase their resilience to economic crisis and sudden downturns. The concern to assess the impact of the crisis in cities, differentiating their resilience capacity, was apparent in 2010 in the URBACT[8] project where the results of an inquiry guided by this approach were published (URBACT 2010).

In Portugal, one of the known examples of the practical application of the paradigm to a regional development plan (still on-going) is the "Alto Minho Development Plan". The second part of the title sets out the strategic aim of this plan, which results in the following formulation: "how to make the Alto Minho a more resilient region" (Mateus 2013). The methodology that shapes the development of this instrument is based on the resilience framework, guided by three purposes: (i) resilience for sustainability; (ii) resilience for cohesion; (iii) resilience for competitiveness.

Another example of the adoption of this conceptual framework is found in the Regional Operational Programme of Central Portugal (2014–2020) (CCDR-Centro 2014). Furthermore, the use of this paradigm is reflected in the focus of the strategy for the Responsible Competitiveness, Structuring, and Resilience (RCSR) model. The authors of the programme aim to create a resilient region, "in the sense of being robust against the context of oscillations, drawing a positive development path that is able to withstand different types of contingencies that may arise at national and international levels, as well as in relation to the good, and less good, moments" (CCDR-Centro 2014, p. 12).

There are several reasons that guide this progressive approach (of theoretical and practical application). Some of these issues are: stopping the domain of armed, civilizational or diplomatic conflicts, subduing the dialectic of development *versus* underdevelopment, and emptying part of the accumulated capital for sustainability; as a result, the crises constitute the backbone of global concerns. Its multiple variants are: environmental (directly or indirectly connected to climate changes), insecurity (especially after 9/11), financial, sovereign debt, economic, institutional, and social

[7]http://www.espon.eu/main/Menu_Projects/Menu_AppliedResearch/ECR2.html

[8]http://urbact.eu/en/news-and-events/view-one/urbact-news/?entryId=5008

(the latter triggered in 2008 via *subprime*). These have redirected a part of the scientific community to scrutinize the aforementioned problems, questioning the paradigm that marks Western development: that is, despite the cyclical oscillations, this paradigm should be driven against a referential (trend) of gradual growth without disruptions.

The impact of Hurricane Katrina and the Al Qaeda's attack (9/11) shifted the focus of part of the research (primarily in the USA) to phenomena/processes that are labeled as "*comebacks*" (Pendall et al. 2010). These two events of global impact led to the increasing awareness of the need to understand the mechanisms that enable territories to recover from traumatic situations.

A window is opened to unpredictability and thus the issue of balance as an untouchable gravitational force is put into question, as are the ability of responsiveness of linear thinking, the cause/effect relationship and the feasibility of the system partitioning to explain parts of a whole. Alternatively, explanations are sought based on the possibility of multiple 'multiple balanced situations and in the integrated functioning of the systems, taking into account the effects of "*feedback*" and its complexity. It is in the convergence of these approaches that the referential of resilience grows (and gains consistency).

Alongside the mounting relevance given to the unpredictability and subjectivity that result from interactions between multiple components of the social systems, there is some unawareness in face of the extension of the interpenetration between environmental systems and the spheres of transformation/human uses of resources. The analytical models, which have been used for observing the complexity brought forth by the approach to multiple meridians of life (of) on the planet we inhabit, are seen as precarious. Given the complexity of all the elements, uncertainty rises.

Uncertainty emerges and expands (Fig. 3.3) whenever the existing knowledge is ambiguous (not fully defined), the contours of the problems are completely unknown (ignorance), and every time risks are internalized in attempts to understand the phenomena and the relationships these establish (possibility of existing intra- and inter-boundary failures) (Rocha 2012).

As she underscores the relevance of the spreading of uncertainty, Davoudi (2012) relates it to a growing sense of unpredictability about the future. In the words of the author, we live in challenging times, plagued and impregnated by high uncertainty. She continues: we are confronted daily with unpredictable elements, overshadowing our ability to anticipate what we will be found around the corner. In the context of urban planning, this constant tension leads to the replacement of the discourse of sustainability by the desire to establish specific prescriptions for resilience. It has constituted an analogous process that led to the gradual focus on the "environment" due to the profusion of the imperative to understand (or meet) "climate changes" (Davoudi 2012). This process is not exempt of doubts, rebukes, and criticisms to which it is necessary to pay attention.

To reduce risks, mitigate the impact, as well as streamline and strengthen responses have been the most commonly proposed mechanisms to deal with, or bypass, crises. However, as Martin-breen and Anderies (2011) suggest the solutions have not been consistent enough to solve the problems they address. As examples of

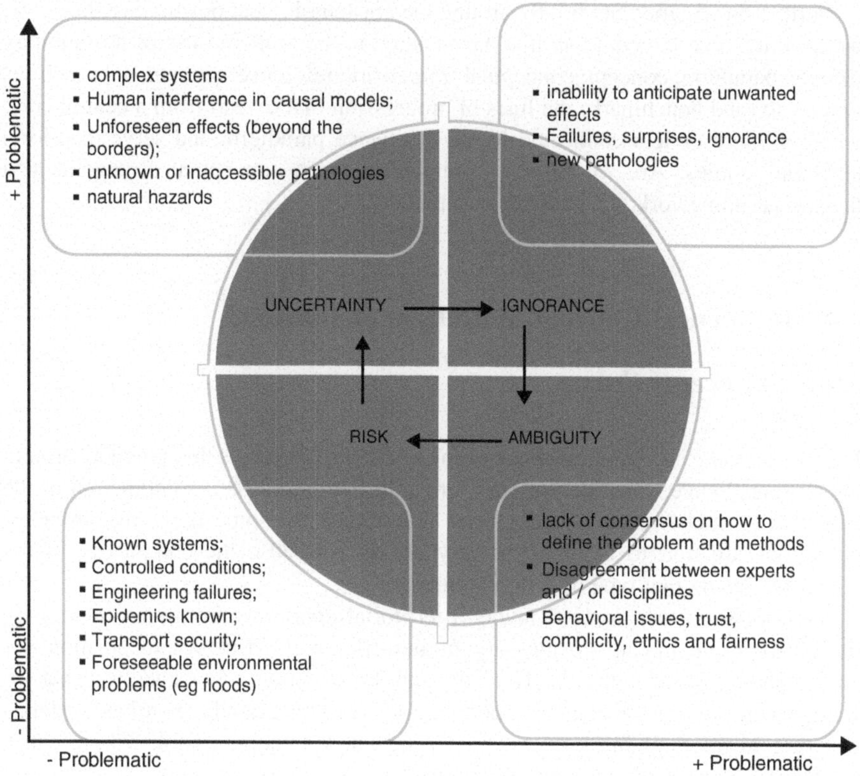

Fig. 3.3 Uncertainty dimensions in Andy Stirling's interpretation. Source: Adapted from Rocha (2012)

this, the authors report the insufficient responses to: chronic poverty, the inconsistency of actions against the concentration of wealth and human potential, the little importance given to global warming, the neglectful way in which growth and concentration of population are dealt with or the increase of resource consumption. In short, in the fragile path taken towards the aim of sustainability.

In this framework, resilience (despite having ancient origins, strengthen as seen above in the long-gone 70s), as a structure of theoretical reflection, gained ground over the last decade, spreading to various sciences, after a period of incubation in engineering, psychology, and ecology.

More recently this theoretical framework has "infected" a plethora of areas from political sciences, management, sociology, history, disaster prevention to urban planning and regional development (Martin-breen and Anderies 2011). However, and in spite of the first image of progress, there is still a strong convergence of the concept in the areas it was originally worked on: psychology and psychiatry (56% of total quotes). Since these are not the views this study focuses on, next we seek to establish how the concept was expanded and unfolded.

Further on we gravitate briefly around the epistemological positioning. For now, we have to solve the equation that has resulted in the scattered use of terminology (theory, paradigm, concept, conceptual framework, reference, rational, approach…) to refer to (and benchmark) the lines of progression that extend from a central core, which is more or less connected by the resilience paradigm, and serve to inform territorial studies. Are we before a plural concept, a conceptual arrangement, a theoretical framework, a paradigm, or a theory?

3.4 Resilience: Concept, Paradigm or Illusion?

3.4.1 Concepts Frame

In order to make a general coverage of the number of concepts that gravitate around the term resilience, some definitions were collected based on a combination of the sources to which we attributed greater importance. In some cases the proposed definitions are not, to begin with, very concise, because the complexity of the concepts hinders that type of formulation.

This sequence of "*flashes*" favors a territorial approach (from an evolutionary perspective) and having no glossary framework, firstly it serves to position our understanding of the concepts. Thus, the path for discussion is opened to debate to what extent we can or cannot undertake the Resilience study of urban systems, inserting this research in a process of asserting a new paradigm. Table 3.3 presents a set of key elements of that set of concepts.

As a transdisciplinary conception, the study of resilience (particularly its socio-environmental dimension) contains a number of weaknesses ranging from the difficulty of accurately defining the territorial scale to the demands of having methodological insights so as to determine the timescale and magnitude of episodes of shock/pressure that territories are subject to. "One of the most intriguing questions of economic geography is to know why some regions are structured to generate self-renewal, while others (they drag into) remain locked in declining frames" (Hassink 2010, p. 1). Here begins the puzzle that the studies of socio-environmental resilience intend to assemble in the broad sense.

3.4.2 Between Paradigm and Illusion

From the attempt to put into the same framework of analysis a set of rules, concepts, that is principles that guide a research can result (when its coherence and cohesion are noticeable) in the configuration of a paradigm. Whoever approaches the task of solving a problem to analyse or try to describe the position of a particular branch of science compared to a precise aspect of reality cannot avoid the need to have to align the research compared to a benchmark. If worthy of recognition by the members of a

Table 3.3 Central Frame of concepts associated with the resilience paradigm

Concept	Definition	Combination of sources
Resilience	The ability that a system displays when absorbing disturbances and reorganizing itself as it triggers changes that allow it to maintain the integrity of its functions, its structure, and the skills to trigger response mechanisms and identity maintenance. The ability to change while maintaining the same identity From the urban systems' point-of-view of, resilience categorizes the competence of a region, urban system, or city to anticipate, prepare for, respond to, and recover from a crisis. It is the ability that urban systems exhibit to prevent blockages (keeping them below the optimal level of development, taking into account their capabilities/resources), support a given path or transit to another that appears to be more profitable Combined effects of: (i) resistance (sensitivity level or the consistency of the reaction of an urban system in a crisis situation); (ii) recovery (speed or extent of recovery that an urban system can trigger); (iii) reorientation (the amplitude of reorientation and adaptation that an urban system mobilizes in response to or in anticipation of a crisis; iv) renovation (consistency of the new development trajectory that was triggered) It cannot be seen as a fixed target, on the contrary, it categorizes a dynamic property, a competence (that is manifested, built or destroyed). It reproduces rhythms of incremental changes. It is more than repeated persistence and diligent adaptability. It includes transformability	Folke et al. (2010); Simmie and Martin (2010); Martin (2012); ESPON (2013)
General resilience	Resilience of any and of all the constituent parts of a system to crises, including those that display a typology not experienced in the past	Folke et al. (2010)
Specific resilience	Resilience "of what and for what?". It applies to a specific part of a system, to a particular control variable, or to a particular kind of crisis/shock.	Folke et al. (2010)
Resilience of the urban systems	Level of changes that an urban system is able to absorb before rearranging itself around a new nucleus of structures and processes. It comprehends the flexibility to capitalize opportunities created by uncertainties and unexpected episodes. Dynamics that enables constant rises in the development trajectory in different scenarios It is not limited to responsiveness, anticipation and recovery capacities (which is linked to the equilibrium of logic) in face of a crisis; it reproduces the scope of persistence in the urban system, the transition to pace (ability to continuously introduce incremental changes) and the degree of transformability (the extension of the system's reconfiguration)	Chelleri (2012); Resilience Aliance (2007); Simmie and Martin (2010)

(continued)

Table 3.3 (continued)

Concept	Definition	Combination of sources
Planning resilience	It aims to equip an urban system with features that allow it to deal with gradual and sudden changes. It implies: (i) adopting a detached dynamic perspective of linear mechanics to return to balanced/stability positions of rational; return to normality; (ii) considering the economic, social and ecological heterogeneity not only attached to forms, but also to processes (functioning) of urban systems. It is based on systemic analyses that allow us to identify the focus of vulnerability, the fittest to admit adaptability and those that require transformability. It is made viable by a systemic perspective, as means are defined by leaving ends opened; it requires some flexibility that will allow the urban system to proceed with incremental adaptations that may benefit from expected and unexpected disturbances. It combines the long-term perspective with immediate action	Eraydin (2013)
Socio-environmental system	The pattern of interactions between ecosystems and communities designed by variable geometry chains that enable "boomerang" effects. The concept emphasizes the perspective that mankind is just one of the many elements present in nature. Their interaction triggers multiple actions and reactions	Folke et al. (2010)
Adaptability	The ability that performers in a system demonstrate to manage resilience, moving the system within an operating base, or proceeding with the transition to a new one. Adaptability may involve: (i) distancing from or approaching a system to a given threshold; (ii) making a threshold easier or more difficult to achieve; (iii) promoting interactions at various levels to avoid or generate resilience at broader scales In socio-environmental systems, adaptability incorporates the learning ability and associates it to experience and knowledge in order to undertake the reorientation requested by external pressures and by internal transformation processes, maintaining the same development path, within a certain operational regime. It includes the ability that the performers demonstrate to enhance resilience and manage change, not crystallizing stages	Walker et al. (2004); Folke et al. (2010)
Adaptive cycle	It describes the four stages through which the complex adaptive systems go through: exploration (r), conservation (k), liberation (Ω) and renewal ($\alpha_\cdot$). An adaptive system can be directed to two contradictory purposes: (i) growth and stability; and (ii) change (ability to assimilate and trigger innovation) and variability (make renewal constant) From the point-of-view of the economy of the urban system, the cycle is structured in two parts of a single loop: on the one hand, there are the emergency, development and stabilization conditions of a given economic structure (growth trajectory between exploration and	Holling (2001); Simmie and Martin (2010)

(continued)

Table 3.3 (continued)

Concept	Definition	Combination of sources
	conservation phases, which show high resilience); on the other, there are factors of crystallization and decline, accumulating capital (resources, material, knowledge), generated in the previous phase; in this phase resilience decreases. The knot in the loop is designed in the opening that this phase of conservation provides in order to create new potential, and new types of activities (creativity and reorganization)	
Panarchy	Interaction dynamics between specific parts of the adaptive cycle. It allows us to understand the complexity of the functioning of ecological and social systems since it comprises the interaction between scales and the various system levels. It admits two-dimensional contamination resulting from the tension between persistence and change. It also results from the interaction between scales (spatial and temporal) upstream and downstream, between forces that contribute to stability and those that leverage changes	Folke et al. (2010); Teigão dos Santos (2009); ESPON (2013)
Transformability	The ability to structure a new system, incorporating into it variable conductive paths that allow for new ways of life, which prefigure new landscapes and will eventually recalibrate stability formulas. When the ecological, economic or social settings (including political) confer unsustainability to the existing operating model (constituting locks), transformability is summoned. Hence, space for regime changes is opened	Walker et al. (2004)
Incremental transformation	Phased introduction of one or more variables (new lifestyles, new urban models) in proximity scales and via "bottom up". While the overall resilience of the system is maintained at a high level, points of change are introduced, nourishing the dynamics of incremental development	Folke et al. (2010)
Regime	Set of possible states that do not endanger the stability (operational generator of prosperity) of an urban system, of a landscape, of a region	Walker and Meyers (2004)
Threshold (*"threshold"*)	Breaking point ("ridgeline") between two operating systems (two basins of attraction) of a system. Border that separates two distinct models of development	Walker and Meyers (2004)
Robustness	Along with resilience, it presupposes the ability to keep the system running in a crisis scenario However, it is related to a rigid system and a specific type of crisis. It implies that the observed system does not unlock certain territorial boundaries and that it is located in a given timescale sheltering it from external disturbances To expand or maintain robustness in a section of a system, or in a particular scale, may involve introducing vulnerabilities in others	Martin-breen and Anderies (2011); Cifdaloz et al. (2010); Salat and Bourdic (2012); Normandin et al. (2009)

(continued)

Table 3.3 (continued)

Concept	Definition	Combination of sources
	The robustness of an urban system can be evaluated using the level of disaggregation it holds and the behavior it manifests when some of its points of support are removed. It collapses instantly or gradually comes into failure	
Vulnerability	Antonym of resilience. However, it requires clarification at all times: 'Vulnerable to what?'. It defines (for example) an economic base exposure to risks of exogenous disturbances resulting from its level of openness (concentration in exports or dependence on imports in strategic factors). It represents the susceptibility to suffer damage before a certain disturbance/crisis/disruption	Martin-breen and Anderies (2011); Briguglio et al. (2008)
Sustainability	Preserves anything, or any function. Normally, it signals a structure whose preservation is desirable, so it can continue and be useful in the future. In some configurations, the path to sustainability may not involve resilience gains (positioning of aversion to risk, performances focused on recovery during a crisis, focus on the capitalization of efficiency). Accepting that disruption of a certain kind and of a certain magnitude cannot be avoided, then sustainability, as a projection of certain frames in a long-term period, requires constant increments of resilience (adaptation with steady gains)	Martin-breen and Anderies (2011)
Equity	It implies social consensus between justice and fairness in the distribution of costs and benefits of a policy, of a program aimed at providing public services. It implies discovering: 'Who benefits from what?' and 'who pays for what?' It is based on the provision of infrastructure and collective equipment as well as universal access to services of general interest. It can be seen in the set of services that are possible to provide, in how they adjust to needs, how they adjust to demand, how they adjust to preferences, how they respond to the (un)willingness to pay for them. Spatial equity (which is a synonym for social and economic equity) refers to the level of accessibility to services and amenities; it depends on its pattern and distribution. This way, various approaches to social cohesion are drawn up Spatial equity can be seen (in a restricted sense) as the accessibility to amenities and to public services, measured as the distance to the points of delivery, or (in a broad sense) including the options of choice (accessibility to and) at work, at teaching institutions, at events/cultural products, of information, of different ranges of services and groups	Barroso et al. (2011); Omer (2006) Truelove (1993); Kunzmann (1998)

(continued)

Table 3.3 (continued)

Concept	Definition	Combination of sources
Accessibility	It has dimensions that extend the direct relation to physical distance. Aspects such as physical handicaps, income, information availability and education deficits influence the accessibility standards. Equity and accessibility are important components in the process of increasing cohesion and resilience	Marques da Costa et al. (2012)
Diversity	In the structure of urban systems, diversity (economic, social, morphological) corresponds to the richness of the *mix* of objects, performers, and structures; it is associated with the potential for creativity and spatial justice statement. As such, diversity can be taken as a target of urban planning Often the subtraction of diversity in urban systems (carried out mainly during the twentieth century) is associated with the dispersion of the population, to the segmentation of the city by functions (removal of industrial activities from the centre, establishment of new cities and new mono-functional centres in a wider region) and an increased socio-spatial segregation. Multi-functionality, diversity and heterogeneity are concepts which are reinforced. Batty et al. (2003), following-up the logic of *"Christallian"*, suggest a measure of diversity, considering the number of activities in certain areas while taking into account the total activities already present in an urban system. Diversity requires minimum levels of redundancy, thus opposing the logic of optimizing efficiency Inseparable from these readings, diversity emerges as a fundamental property of resilience and sustainability, indicating the existence of multiple forms, environments, activities and communities as well as ensuring greater capital to generate opportunities and circumscribing dependence in face of restricted nucleus seen as drivers of progress. Diversity reflects the variety, or the ability that a community has to generate resources (contact networks, social bonds, reciprocity, coordination, and cooperation...) and knowledge to build resilient urban systems (cities or regions) Maintaining conditions to generate on-going incremental change is a prerequisite to design the structure of urban systems long-term and ensure its diversity	Fainstein (2005); Batty et al. (2003); Fiksel (2003); Gotham and Campanella (2010); Clifton (2010); Perrings (2006); Holling (1996)

Source: combination of sources

scientific community (or sub-community), this referential may assume the paradigm status. This does not mean that the task of selecting a set of rules that can embody a tradition of a particular science is a simple undertaking. Instead, it is more likely to become a source of continuing frustration. It must be said that the pursuit of this consistency is not an end in itself. The lack of interpretative standardization, or the absence of unanimity in the code of guiding rules, does not prevent a paradigm from marking a research.

In fact, the existence of a paradigm does not have as a prerequisite the existence of a closed set of rules (Kuhn 2006). From Kuhn's interpretation results in the argument that scientists conduct their research using models transmitted to them along their training course and the literature framework they collect. Transparency about the paradigm from which the guiding model of analysis of their work emanates may not result from this (Fig. 3.4), though. Therefore, it is not for this reason that research results are weakened.

Research does not necessarily depend upon a cohesive set of rules or assumptions. This set owes coherence to the attempt to explain a concrete section of phenomena and to the specific applications that might result from such process (Kuhn 2006). These established rules (that define a paradigm) are widely accepted when the explanations for the problems that motivated their existence are consensual and fade away in the phases in which insecurity arises in these explanations. Kuhn (2006) calls this period where anomalies appear (crises, insecurity, inability to produce the expected results) "pre-paradigmatic", which are the precursors of "scientific revolutions" (large and small). A small revolution can be the discovery of a new way to explain a circumscribed phenomenon. It is this way (i.e., in pre-pragmatic periods) that discussions take place about methods, problems, and consistency of patterns of existing answers, which are set-aside in "periods of normal science". "Failure of existing rules is the prelude to a search for new rules" (Kuhn 2006, p. 95).

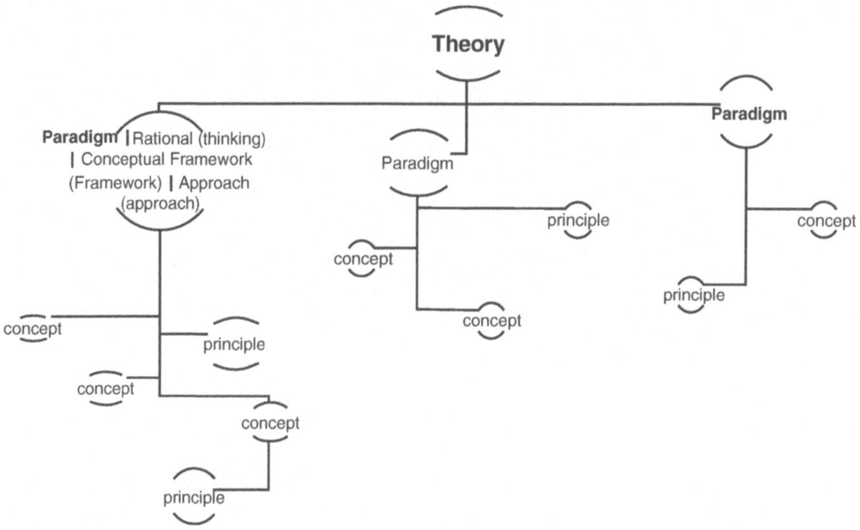

Fig. 3.4 Schematic representation of epistemological organization. Source: The author

The paradigm of socio-environmental resilience (which we will see if it can be regarded as such) is a promising concept because it both demands for thematic integration and looks for answers for the underlying issue that there are territories that when threatened by critical situations are able to trigger responses that reposition these regions in order to improve the quality of life in their communities whereas others stabilize, recede or delay their ability to react in time. However, especially when considering the operability of the research, it shows a number of weaknesses that must be circumscribed as of now. For that reason, next, we will present some of these critical points.

The difficulty of thematic integration is perhaps one of the first critical points to be faced in this area of research. This situation requires large research teams to gather in order to cover the different components included in the functioning of a territory. These issues may extend from the effects of climate changes to the role of agriculture going through the analysis of human capital, but always keeping in mind that it is crucial to meet conditions that enable the creation, reinforcement or management of resilience.

Generally, it is necessary to develop efforts to study the complexity of the systems to the extent that if we agree with Zimmerer (1994), Adger (2000), Holling (2001) and Limburg et al. (2002), among many others, we are led to conclude that ecological systems share many features with social systems. In fact, both are complex structures divided into parts that are related by dynamic processes. Therefore, they only benefit from being treated simultaneously. As open and dynamic systems (where it is not easy to separate what is incorporated and what is external to the analysis and in which various types of interactions occur), complexity increases exponentially.

The difficulty of developing compatible time, space and processes scales is the second group of weaknesses. The territorial scale consists of the exact extent of the urban system where it is possible to distinguish without great effort the endogenous components from those which are exogenous. Knowing that they both reproduce interactions in the system, the responsibility of evaluating the intervention of performers and external factors is likewise critical as is measuring the effect of their decisions and interventions in the resilience capacity of a specific area/region. Yet the assessment of the scope and nature of the processes is still in question. What is the scale of the phenomenon that triggers the vulnerabilities of the territories to the different risks (loss of human lives, loss of adaptability, loss of ability to generate quality of life)? The difficulty to consider the role of governance in the resilience of territories both in terms of the social capital generated and in terms of how they are managed and how both transfer effects on their resilience capacity is another area where the analysis effort is still incipient.

Understanding the consistency of the criticism made to the resilience approach is one of the starting points that will enable us to assume more effectively the advantages and disadvantages of the research channels. In this context, it is important not to put aside the argumentative skepticism of authors such as Hassink (2010). He warns us about the attraction that human geography, in particular, the economic geography (Markusen 1999), exhibits before new concepts, which in many cases are void in what concerns the strengths of the research they are based on.

Once again we refer to Swanstrom (2008) to underscore the point intended for discussion here. Since we agree with the author, we have to recognize that given their state of maturity studies (particularly those whose object are urban systems) which adopt the framework of resiliency occur on a conceptual framework structure that despite being more than a metaphor are still less than a theory. As the best hypothesis, the author defends a conceptual framework to help us to think about cities, regions, or urban systems in different ways. Immediately, Swanstrom makes it impossible to reject the dynamic and holistic perspectives.

However, this reading is not consensual. In the context of cities and bearing in mind the risk management that affects them (environmental, technological, financial, economic, social), resilience represents a new paradigm to face the urbanization process, influencing how to manage exposure to vulnerability or strengthen mechanisms to enable triggering responses. These principles are also valid to position urban planning in a general way (Jha et al. 2013). The path to finding answers to the increasing occurrence of extreme events (in the environmental and technological levels) have recently called upon researchers to seek new paradigms that place an emphasis on resilience for the sustainability of communities and the resilience of societies faced with multiple unpredictability (Tobin 1999).

The use of the paradigm idea to catalog the proposed development that underlies resilience seems to be recurrent. The need for a paradigm change is considered when facing the challenge of designing a green economy (gradually removing carbon from its composition): how to confer economic resilience, particularly for vulnerable communities, such as eradicating poverty, how to articulate the sustainability of the production and consumption circuits, and how to keep increments of development in a world where it is possible to restrict carbon emissions (Panel of High Level of the Secretary-General of the United Nations on Global Sustainability 2012). We uphold the idea that the ability to maintain territorial systems within the limits of the running regulations calls for a set of prerequisites (adaptability, portability, robustness, flexibility, redundancy, modularity), which contribute to its resilience while working as an aggregator nucleus of these principles.

In a paradigm, the arrangements and rearrangements that structure a development model are defined. The current paradigm gravitates around an attractor (economic growth) that has gradually lost strength, making the transition to a new attraction base a desirable (if not inevitable) scenario (Folke et al. 2010). From this point-of-view, it will be indispensable for the new base of attraction to be able to set its unifying nucleus through which the transition to a new paradigm can be directed. We would therefore have to accurately identify which motor of the current paradigm (the plan directed towards growth (Rieniets 2009)) is unsustainable and hence tends to crises and to collapse (William 2010) so that the conceptual framework of substitution (development, sustainable development, resilient development for sustainability) gradually occupies the space created by the receding one. This sphere of governance performance has a central role in the transition process from an urban paradigm displaced from the cities', regions' or countries' productivity to another type of model, more able to meet prosperity benefits distribution (existing or to be generated) (UN-HABITAT 2012).

Salat and Bourdic (2012) have another type of approach that refers to the resilience of urban space and searches for criteria to support a self-organization facilitation paradigm, a spontaneity logic, and informality in urban expansion. Synnott and Griffiths (2012) use the term "*resilience theory*", stating that through it we have an opportunity to consider a new transformation paradigm, where the city and the systems in which it operates are guided by different rules. The city is itself a paradigm of the challenges that society faces. If not for other reasons, because it is where the major pathologies of our time crystallize (Hickey 2007).

A paradigm change will only be visible in retrospective. It depends on its success in the future since it does not just work by considering what has failed in the past. As we observed, classical approaches do not seem to be working. However, the resilience thought framework provides elements capable of creating new answers. Although still untried, these possibilities may be the best chance that Mankind has at its disposal (Martin-breen and Anderies 2011).

Anderies et al. (2006) and Chelleri (2012), among other authors, prefer to use the term "*framework*" which we can take to mean a conceptual framework, framing, or an integrative platform. The possibility of the term being accepted as a theory is withdrawn, advocating that it is more appropriate to associate it with the idea of a conceptual framework to systematize a line of thought that aims to analyse the dynamics of socio-environmental systems. It allows us to gather elements from several theories in many studies based on resilience (developed in the economy, the ecology, the dynamics of systems), building bridges that solidify the analytical skills, by means of its joint understanding.

This uncertainty, coupled with the weaknesses identified above, often brings out (Davoudi 2012; Rose 2009; Fernandes and Chamusca 2012) an alert to the risk of there being an on-going buzzword consolidation ("*buzzword*") whose sense, because it is unstable, can become void. Planning has a base for this type of promising approaches that often end up becoming "*slogans*" when processed and imported into technical, political or media language. Questions like: 'can its improper and indiscriminate use serve to defend the indefensible?'; 'Can the paradigm of resilience actually be promising for practical planning?'; 'What opportunities are opened and what limitations arise from its transfer from ecology to social systems, from these to urban systems, and from there to planning?' (Davoudi 2012). The risk of becoming a buzzword increases with its use in technical and scientific publications titles simply because it is fashionable, but without any corresponding content. Rose (2009) suggests as an example of this the study "*Resilient City: The Economic Impact of 9/11*" (Chernick 2005).

Fernandes and Chamusca (2012) are an example in the opposite direction. These authors support the validity of the paradigm, which in this case is the interpretation of changes in commercial spaces. For these authors, the paradigm's grasp is beyond the process of creating another buzzword. They give it great importance to manage crisis situations (such as we are experiencing now) since it can lead to the introduction of innovations in policies rather than mere additions. The paradigm will also have a crucial role in fighting proposals that often come across the public space and that suggest urban revolutions, new cities, but forget existing structures, neglecting their ability to be able to be absorbed into the system, reacting, adapting, cooperating and promoting continuous change (Fernandes and Chamusca 2012, p. 2).

Having said this, we return to Swanstrom's interpretation (2008), which states that resilience studies are positioned between metaphor (we add: the concept, the principles) and theory. Hence, one must accept that one works in the midst of an on-going defining process of a paradigm that is still open but is already strong enough to be considered as such. Bearing in mind that existing explanations are weakened (see the previous section) and the "community" working around this rationale of concepts, principles, and rules, albeit dispersed, is (as can be seen by what we wrote at the beginning of this point) clearly increasing are two strong reasons that justify Swanstrom's and our own reading.

3.5 Conclusion

Planning resilience implies a dichotomous point-of-view that places the analysis of the transition processes in its core. It also implies that we evaluate the paradigm's changing mechanisms to a new vision of sustainable development where resilience is seen as normative. Hence we reconnect the concept to its origin, fostering greater understanding of how to inject sustainability in the urban systems' growing complexity. It is in this base that the resilience paradigm thrives and it is best to consider its evolutionary aspect.

With this purpose in mind, we searched for the etymological roots of the concept; its progress from what we can call "science of resilience" was identified so that next we could discuss more clearly its admission into a conceptual, paradigmatic space, without neglecting the fields that expose the continued weaknesses of this approach. The evolution registered from the interpretation made by the engineering field to its seizure by ecology, and the leap from there to territorial systems, opened possibilities and allowed for the configuration of a conceptual framework consistent enough to challenge development paradigms. Positioning the territories as gauges, it is possible to identify features that minimize their weaknesses and others that amplify their sturdiness. The notions of adaptability and above all the embedded formulation in the ideas of an adaptive cycle and panarchy help to organize this broad and dispersed conceptual framework. Hereafter, it drifts into formulations that set the idea of region, city, and resilient community.

A region, such as a city or a resilient community, develops a planning model based on gradual and continuous changes, where opportunities are expanded for all age and social groups; a connectivity network is established and consolidated, learning conditions are internalized (in its socio-economic matrix) which avoid or innovate from mistakes and tension scenarios. In addition, the model positions itself to integrate territorial systems that enhance interactions at multiple scales. At the centre of the performances is the need to facilitate a resilience culture in the state, organization and individual spheres. A resilient community maintains, regains, or establishes favorable results over time (regardless of the crises episodes) as well as triggers, without any interruption, productive changes on everyday life scenarios.

References

Adger, N., Hughes, T., Folke, C., Carpenter, S., & Rockström, J. (2005). Social-ecological resilience to coastal disasters. *Science, 309*(5737), 1–6.

Adger, W. N. (2000). Social and ecological resilience: Are they related? *Progress in Human Geography, 24*(3), 347–364.

Alexander, D. (2013). Resilience and disaster risk reduction: An etymological journey. *Natural Hazards and Earth System Science, 13*(11), 2707–2716.

Anderies, J. M., Walker, B. H., & Kinzig, A. P. (2006). Fifteen weddings and a funeral: Case studies and resilience-based management. *Ecology and Society, 11*(1).

Bacon, F. (1625). *Sylva sylvarum, or of natural history in ten centuries*. London: Lee, W.

Barroso, S., Costa, E. M., & Andrade, I. (2011). Desenvolvimento urbano sustentável em portugal: uma abordagem integrada. ECORYS UK limited, for the European Commission, Directorate-General for Regional Policy.

Batabyal, A. A. (1998). The concept of resilience: Retrospect and prospect. *Environment and Development Economics, 3*, 235–239.

Batty, M., Besussi, E., Maat, K., & Harts, J. J. (2003). *Representing multifunctional cities: Density and diversity in space and time* (Working papers series, paper 71). London: Centre for Advanced Spatial Analysis University College.

Bell, R. (1839). *Eminent literary and scientific men. English poets, Dionysius Lardner's cabinet of biography series* (Vol. 2). London: Longman.

Béné, C., Wood, R. G., Newsham, A., & Davies, M. (2012). *Resilience: New utopia or new tyranny? Reflection about the potentials and limits of the concept of resilience in relation to vulnerability reduction programmes* (Working paper 405). London: Institute of Development Studies.

Berkes, F. (2007). Understanding uncertainty and reducing vulnerability: Lessons from resilience thinking. *Natural Hazards, 41*(2), 283–295.

Berkes, F., & Folke, C. (Eds.). (1998). *Linking social and ecological systems: Management practices and social mechanisms for building resilience*. New York: Cambridge University Press.

Brand, F. S., & Jax, K. (2007). Focusing the meaning(s) of resilience: Resilience as a descriptive concept and a boundary object. *Ecology and Society, 12*(1).

Briguglio, L., Cordina, G., Farrugia, N., & Vella, S. (2008). Economic vulnerability and resilience concepts and measurements. *Oxford Development Studies, 37*(3), 229–247.

Brock, J., Göran, K., & Mäler, M. (2002). Resilience and sustainability: The economic analysis of non-linear dynamic systems, 261–292. Chapter 10. In L. Gunderson, D. Ludwig, & C. Holling (Eds.), *Panarchy: Understanding transformations in human and natural systems*. Washington DC: Island Press.

Carpenter, S., Walker, B., Anderies, M., & Abel, N. (2001). From metaphor to measurement: Resilience of what to what? *Ecosystems, 4*(8), 765–781.

Chelleri, L. (2012). From the «Resilient City» to urban resilience. A review essay on understanding and integrating the resilience perspective for urban systems. *Documents d'Anàlisi Geogràfica, 58*, 287–306.

Chernick, H. (Ed.). (2005). *Resilient city: The economic impacts of 9/11*. New York: Russell Sage Foundation.

Cifdaloz, O., Regmi, A., Anderies, J. M., & Rodriguez, A. A. (2010). Robustness, vulnerability, and adaptive capacity in small-scale social-ecological systems: The Pumpa irrigation system in Nepal. *Ecology and Society, 15*(3), 39.

Clifton, D. (2010). Progressing a sustainable-world: A socio-ecological resilience perspective. *Journal of Sustainable Development, 3*(4), 74–96.

Collier, M., Nedović-Budić, Z., Aerts, H., Connop, S., Foley, D., Foley, K., & Verburg, H. (2013). Transitioning to resilience and sustainability in urban communities. *Cities, 32*, 21–38.

Comissão de Coordenação e Desenvolvimento Regional do Centro. (2014). Programa Operacional Regional do Centro de Portugal (2014–2020). Coimbra.

Cumming, G., Barnes, G., Perz, S., Schmink, M., Sieving, K., Southworth, J., et al. (2005). An exploratory framework for the empirical measurement of resilience. *Ecosystems, 8*(8), 975–987.

Davoudi, S. (2012). Resilience: A bridging concept or a dead end? *Planning Theory & Practice, 13* (2), 299–333.

Djalate, A. (2011). Resilience and disaster risk reduction: An etymological journey. *Natural Hazards and Earth System Sciences, 1*, 1257–1284.

Eraydin, A. (2013). Resilience thinking for planning. In A. Eraydin & T. Taşan-Kok (Eds.), *Resilience thinking in urban planning* (pp. 17–38). Ankara: Ayda Erayd.

ESPON. (2013). *Economic crisis: Resilience of regions, inception report*. Luxembourg City: ESPON/Cardiff University.

Fainstein, S. (2005). Cities and diversity: Should we want it? Can we plan for it? *Urban Affairs Review, 41*(1), 3–19.

Fernandes, J., & Chamusca, P. (2012). Urban policies, planning and retail resilience. *Cities*, 170–177.

Fiksel, J. (2003). Designing resilient, sustainable systems. *Environmental Science & Technology, 37*(23), 5330–5339.

Folke, C. (2006). Resilience: The emergence of a perspective for social–ecological systems analyses. *Global Environmental Change, 16*(3), 253–267.

Folke, C., Carpenter, S., Elmqvist, T., Gunderson, L., Holling, C. S., & Walker, B. (2002). Resilience and sustainable development: Building adaptive capacity in a world of transformations. *Ambio, 31*(5), 437–440.

Folke, C., Carpenter, S., Walker, B., Scheffer, M., Elmqvist, T., Gunderson, L., & Holling, C. S. (2004). Regime shifts, resilience, and biodiversity in ecosystem management. *Annual Review of Ecology, Evolution, and Systematics, 35*, 557–581.

Folke, C., Carpenter, S. R., Walker, B., Scheffer, M., Chapin, T., & Rockström, J. (2010). Resilience thinking: Integrating resilience, adaptability and transformability. *Ecology and Society, 15*(4), 20.

Goldstein, S., & Brooks, R. B. (Eds.). (2006). *Handbook of resilience in children*. New York: Springer Science & Business Media.

Gotham, K. F., & Campanella, R. (2010). Toward a research agenda on transformative resilience: Challenges and opportunities for post-trauma urban ecosystems. *Critical Planning, 17*, 9–23.

Gross, J. S. (2008). *Sustainability versus resilience: What is the global urban future and can we plan for change*? (Discussion paper). New York: International Center for Scholars and the Fetzer Institute.

Gunderson, L. H., & Holling, C. S. (Eds.). (2002). *Panarchy: Understanding transformations in human and natural systems*. Washington DC: Island Press.

Hassink, R. (2010). Regional resilience: A promising concept to explain differences in regional economic adaptability? *Cambridge Journal of Regions, Economy and Society, 3*(1), 45–58.

Hickey, M. (Ed.). (2007). The city of tomorrow – proceedings of the international colloquium on the future of cities. In *Proceedings from the international colloquium on the future of cities*. Bruxelles: Territorial Development Agency.

Holling, C. S. (1973). Resilience and stability of ecological systems. *Annual Review of Ecology and Systematics, 4*, 1–23.

Holling, C. S. (1996). Engineering resilience versus ecological resilience. *Engineering Within Ecological Constraints, 31*, 31–43.
Holling, C. S. (2001). Understanding the complexity of economic, ecological, and social systems. *Ecosystems, 4*(5), 390–405.
Jha, A., Miner, T., & Stanton-Geddes, Z. (2013). Building urban resilience principles, tools, and practice. In T. W. Miner, Z. Stanton-Geddes, & A. K. Jha (Eds.), *Directions in development*. Washington DC: World Bank.
Kuhn, T. S. (2006). In tradução B. V. Doeira e N. Boeira, *A estrutura das revoluções científicas* (9. ed., p. 1970). São Paulo: Perspectiva.
Kunzmann, K. R. (1998). Planning for spatial equity in Europe planning for spatial equity in Europe. *International Planning Studies, 3*(1), 101–120.
Limburg, K. E., O'Neill, R. V., Costanza, R., & Farber, S. (2002). Complex systems and valuation. *Ecological Economics, 41*(SPECIAL ISSUE: The dynamics and value of ecosystem services: Integrating economic and ecological perspectives), 409–420.
Markusen, A. (1999). Fuzzy concepts, scanty evidence, policy distance: The case for rigour and policy relevance in critical regional studies. *Regional Studies, 33*(9), 869–884.
Marques da Costa, E., Palma, P., Rauhut, D., Humer, A., & Constantin, V. (2012). Indicators of services of general interest in EU regional context: Between the need to measure and the lacking of their meaning. In *52nd annual congress of the European regional science association international* (pp. 1–22). Bratislava.
Martin, R. (2012). Regional economic resilience, hysteresis and recessionary shocks. *Journal of Economic Geography, 12*(1), 1–32.
Mateus, A. (2013). Plano de Desenvolvimento do Alto Minho: Como tornar o Alto Minho uma região mais resiliente, *Alto Minho desafio* 2020 (p. 14). Arcos de Valdevez: Comunidade Intermunicipal do Alto Minho.
Normandin, J. M., Therrien, M. C., & Tanguay, G. A. (2009). City strength in times of turbulence: Strategic resilience indicators. In *Joint conference on city futures* (pp. 1–22). Madrid.
Omer, I. (2006). Evaluating accessibility using house-level data: A spatial equity perspective. *Computer, Environment and Urban Systems, 30*, 254–274.
Ott, K., & Döring, R. (2004). *Theorie und Praxis starker Nachhaltigkeit*. Marburg: Metropolis.
Panel of High Level of the Secretary-General of the United Nations on Global Sustainability. (2012). *Resilient people, resilient planet: A future worth choosing*. New York: United Nations Secretary-General's High-level Panel on Global Sustainability.
Pendall, R., Foster, K., & Cowell, M. (2010). Resilience and regions: Building understanding of the metaphor. *Cambridge Journal of Regions, Economy and Society, 3*(1), 71–84.
Perrings, C. (2006). Resilience and sustainable development. *Environment and Development Economics, 11*(4), 417–427.
Pickett, S. T. A., Cadenasso, M. L., & Grove, J. M. (2004). Resilient cities: Meaning, models, and metaphor for integrating the ecological, socio-economic, and planning realms. *Landscape and Urban Planning, 69*(4), 369–384.
Pinho, P. (coord), Oliveira, V., Cruz, S. S., Sousa, S., Martins, A., Morgado, S., Costa, J., & Dias, L. (2008). *SUPERCITIES sustainable land use policies for resilient cities* (pp. 1–402). Porto.
Rankine, W. J. M. (1867). *A manual of applied mechanics*. London: Charles Griffin and Co.
Rieniets, T. (2009). Shrinking cities: Causes and effects of urban population losses in the twentieth century. *Nature and Culture, 4*(3), 231–254.
Rocha, F. (2012). *Sistemas Complexos, Modelação e Geosimulação da Evolução de Padrões de Uso e Ocupação do Solo, Dissertação de Doutoramento, Instituto de Geografia e Ordenamento do*. Território: Universidade de Lisboa.
Rose, A. (2009). *Economic resilience to disasters* (CARRI research report 8) (pp. 1–58). Oak Ridge, TN: Community and Regional Resilience Institute.
Salat, S., & Bourdic, L. (2012). Systemic resilience of complex urban systems. *TeMA Journal of Land Use Mobility and Environment, 2*, 55–68.

Schroll, H., Thorn, P., Kjærgård, B., & Co-editors, T. E. S. (2009). Resilience is more than an elastic jump. *Te Journal of Transdisciplinary Environmental Studies, 8*(1), 1–2.

Simmie, J., & Martin, R. (2010). The economic resilience of regions: Towards an evolutionary approach. *Cambridge Journal of Regions, Economy and Society, 3*(1), 27–43.

Swanstrom, T. (2008). *Regional resilience: A critical examination of the ecological framework* (IURD working paper series) (pp. 1–34). Berkeley, CA: Institute of Urban and Regional Development UC.

Synnott, E., & Griffiths, D. (2012). *Strengthening urban resilience* (pp. 1–17).

Teigão dos Santos, F. (2009). Territórios Resilientes enquanto Orientação de Planeamento. *Prospectiva E Planeamento, 16*, 13–28.

Thapa, S., Marshall, F., & Stagl, S. (2010). *Understanding peri-urban sustainability: The role of the resilience approach*. Brighton: STEPS Centre.

Tobin, G. (1999). Sustainability and community resilience: The holy grail of hazards planning? *Environmental Hazards, 1*(1), 13–25.

Tomes, R. (1857). *The Americans in Japan, an abridgment of the government narrative of the U.S. expedition to Japan under Commodore Perry*. New York: D. Appleton.

Truelove, M. (1993). Measurement of spatial equity. *Environment and Planning, 11*, 19–34.

UN-HABITAT. (2012). *State of the world's cities 2012/2013: Prosperity of cities* (United nations human settlements programme) (pp. 1–152). London.

UNISDR. (2012). *Making cities resilient: My city is getting ready!* (Disaster resilient cities, schools and hospitals) (pp. 1–10).

URBACT. (2010). *URBACT cities facing the crisis responses*. Saint-Denis La Plaine: European Union, European Regional, Development Fund.

Walker and Meyers. (2004). Resilience, adaptability and transformability in social-ecological systems. *Ecology and Society, 9*(2)

Walker, B., Carpenter, S., Anderies, J., Abel, N., Cumming, G., Janssen, M., et al. (2002). Resilience management in social-ecological systems: A working hypothesis for a participatory approach. *Conservation Ecology, 6*(1), 14.

Walker, B., Holling, C. S., Carpenter, S. R., & Kinzig, A. (2004). Resilience, adaptability and transformability in social-ecological systems. *Ecology and Society, 9*(2), 5.

Walker, B., & Salt, D. (2006). *Resilience thinking, sustaining ecosystems and people in a changing world*. Washington DC: Island Press.

William, E. (2010). Thinking "Resilience". In *Post carbon reader: Managing the 21st century's sustainability crises*. Healdsburg, CA: Watershed Media.

Xu, L., & Marinova, D. (2013). Resilience thinking: A bibliometric analysis of socio-ecological research. *Scientometrics, 96*(3), 911–927.

Zimmerer, K. (1994). Human geogography and the "New ecology": The prospect and promise of integration. *Annals of the Association of American Geographers, 84*, 108–125.

Online Document

Martin-breen, P., & Anderies, J. M. (2011). *Resilience: A literature review*. Accessed July 28, 2012, from https://opendocs.ids.ac.uk/opendocs/bitstream/handle/123456789/3692/Bellagio-Rockefeller%20bp.pdf?sequence=1&isAllowed=y

Resilience Aliance. (2007). *A research prospectus for urban resilience*. Acessed July 29, 2012, from http://citiesforpeople.ca/wp-content/uploads/2014/02/urbanresilienceresearchprospectusv7feb07.pdf

Part II
The Multilevel Aspects of Resilience

Chapter 4
Economic Crisis, Turbulence and the Resilience of Innovation: Insights from the Atlantic Maritime Cluster

Hugo Pinto, Elvira Uyarra, Mercedes Bleda, Carla Nogueira, and Helena Almeida

4.1 Introduction

The recent economic turbulence has demonstrated that countries, regions and even firms have different capacity for coping with external shocks. Many fail and are damaged, and ultimately devastated by the impacts of the crisis. On the other hand, some of them are able to resist the shocks, adapt quickly and recover their trajectories, and in some cases even generate new growth trajectories using their relative advantages to deal with turbulent environments (Archibugi et al. 2013; Makkonen 2013; Paunov 2012).

H. Pinto (✉)
Centre for Social Studies, University of Coimbra, Coimbra, Portugal

CIEO – Research Centre for Spatial and Organizational Dynamics, University of Algarve, Faro, Portugal

Faculty of Economics, University of Algarve, Faro, Portugal
e-mail: hpinto@ces.uc.pt

E. Uyarra
Manchester Institute of Innovation Research (MIoIR), Alliance Manchester Business School, Manchester, UK

M. Bleda
Manchester Institute of Innovation Research (MIoIR), Alliance Manchester Business School, University of Manchester, Manchester, UK

C. Nogueira
CIEO – Research Centre for Spatial and Organizational Dynamics, University of Algarve, Faro, Portugal

H. Almeida
CIEO – Research Centre for Spatial and Organizational Dynamics, University of Algarve, Faro, Portugal

Faculty of Economics, University of Algarve, Faro, Portugal

H. Pinto et al. (eds.), *Resilience and Regional Dynamics*, Advances in Spatial Science, https://doi.org/10.1007/978-3-319-95135-5_4

This capacity to respond to shocks and disruptions is the target of increasing policy and academic attention. In regional studies, much emphasis has been placed on the concept of 'resilience' as a capacity of complex adaptive systems to deal with internal shocks and external disruptions (Boschma 2015; Simmie and Martin 2010). Resilience in the study of territorial socio-economic systems has abandoned an engineering and ecological perspective to become an evolutionary concept, focused on the processes of selection, survival, and adaptation, as well as on the adaptability of different types of systems to build new dynamic trajectories by overcoming locks-in and path dependencies (Martin and Sunley 2015). Innovation, knowledge production and exchange, are key contributors to resilience, by creating a variety of opportunities to deal with the challenges that organizations, firms, regions and countries face in highly dynamic and turbulent environments (Simmie 2014). The perception of innovation as a procyclical activity which follows the trend of macroeconomic variables such as GDP and investment, is contested by studies showing firms continuing or increasing their innovative efforts despite the economic downturn (Frenz and Prevezer 2012).

During the years of the economic crisis, innovation and new knowledge creation within systems of innovation, in particular the extent to which new knowledge is generated and diffused across the relevant actors, allowed socioeconomic systems to generate variety and adapt to change (Boschma 2015). Since resilience is often referred to as an attribute of a specific system, some authors suggest that innovation systems, in particular regional innovation systems, are good candidates as a unit of analysis for this capacity (Pinto and Pereira 2014).

In this chapter, we are suggesting a new approach. That resilience is not only seen as an attribute of systems but also as an attribute of the innovation process. 'Resilience of innovation' thus refers to the capacity of an innovation process to maintain or accelerate its functions when facing an internal disruption and/or an external shock. Resilience of innovation, as a complex phenomenon, is a multi-level characteristic that applies to systems at the macro-level, i.e. to individual countries and regions, at the meso-level, in particular focusing networks, clusters and regional innovation systems, and at micro-level of organizations, that is to innovation actors such as firms, universities and other public research organizations, and innovation governance bodies.

As we have already indicated, the focus of this chapter is on the resilience of innovation at the organizational level. Inspired by complex adaptive innovation systems (Cooke 2013) and evolutionary ideas (Boschma and Martin 2010), our goal is the identification of factors that are important to encourage the process of new knowledge generation and exchange which underpins the resilience of innovation processes at the level of the organization. For this purpose, we focus on organizations within a particular regional innovation cluster: the maritime cluster in the European Atlantic Area. In particular, we centre on organizations in this cluster that experienced an increased or unchanged demand for innovation and knowledge-based services, our interest is on those organizations that showed resilience in front of the disturbances consequence of the economic downturn.

For our analysis, we draw upon the results of a survey on knowledge exchange and innovation, which was built in 2014 to detect and assess the specific knowledge needs of entities engaged in maritime cluster in the Atlantic Area, as well as to investigate the provision of innovation and knowledge related exchange services in the sector.[1]

We provide parametric and non-parametric evidence of the differences in the provision and utilisation of these services and of the main factors that influence the resilience of innovation of organizations within the cluster, as well as, several suggestions for innovation policies that can be derived from the analysis.

4.2 Regional Resilience and the Dynamics of Innovation

The concept of resilience is associated to an increase of economic, political and environmental risks and the lack of emergent processes in post-industrial society that have accentuated economic and social inequalities in the regions (Davoudi et al. 2012). The interest in the regions' resilience emerged from a general feeling of uncertainty and insecurity and the search for solutions for adaptation and survival in response to a complex and diverse set of external shocks, including the financial crises. The intersection of the economy with the environment has increased the sense of vulnerability and, therefore, has stimulated the search for new ways to understand the adaptive capacity of regions (Alexander 2013).

Simmie and Martin (2010: 28) defined resilience as an "... 'adaptive ability' to the differential ability of a region or local firms' to adapt to changes and shocks in the competitive market, technological policy and related conditions that the evolutionary dynamics and trajectories of that regional or local economy over time".

To date, work on resilient regions has focused more on conceptual and empirical analysis from high performing regions. An established fact today is that innovation is an essential foundation for resilience and effective social and economic development (Hamdouch and Depret 2012).

The increased focus on regions as the best geographical scale for a knowledge economy points to the importance of geographical proximity and regional resources in stimulating the innovation capability and competitiveness of firms (Cooke and Leydesdorff 2006). The regional innovation narrative is largely based on success stories of specific industrial agglomerations or regional networks of SMEs and industrial clusters (Tödtling and Trippl 2005). In many cases, learning and knowledge transfer are highly localised (Boschma 2005). It is recognized that important elements of the process of innovation become regionalized (Howells 2005). One of the reasons is because innovation occurs in a specific institutional, political and social context (Rodríguez-Pose 2013).

The general perception of the region as the main locus for economic interaction and innovation brought relevance to the notion of "regional innovation system". The

[1] The survey was part of the European project KIMERAA (available at www.kimeraa.eu) aimed at developing economic niches of excellence through the creation of strong linkages between firms and science organizations within the marine sciences and maritime activities.

rise in the popularity of the concept of regional innovation systems has been in part driven by the increased intensity of international competition in a globalising economy, the apparent shortcomings of traditional regional development models and policies, and the emergence of successful clusters of firms and industries in many regions around the world (Uyarra and Flanagan 2012).

As suggested by Doloreux (2002) the concept of regional innovation systems is difficult to delimitate but usually it is understood as a set of private and public interests, institutions and organizations, their relationships that are encouraging the generation, use and dissemination of knowledge. This set produces pervasive and systemic effects that encourage firms within a regional context to develop specific forms of capital that reinforce regional innovation capability and competitiveness (Gertler 2003). RISs are premised on innovation being a geographical process and innovation capabilities being sustained through regional communities that share common knowledge bases (Asheim et al. 2005). The RIS literature supported this argument and showed that firms' innovative activity is based on localized resources such as a specialized labour market and labour force, subcontractor and supplier systems, local learning processes and spillover effects, local traditions for co-operation and entrepreneurial attitude, supporting agencies and organizations and the presence of customers and users (Asheim et al. 2011). On other hand, innovation can occur more easily through organized proximity, regardless of the geographical concentration (Torre and Rallet 2005). The "organized" characteristic refers to the arranged nature of human activities and not to the fact that one may belong to one organisation in particular. It goes beyond the mere cognitive dimensions resting in two key aspects: the belonging and the similarity. Clusters are seen as relevant in contributing to "related variety" (Frenken et al. 2007) and helping the economic interactions between regional actors.

Clusters stimulate sectoral specialisation, cognitive and geographical proximity, competition and cooperation, leading to spillovers and synergies within a regional innovation system. Innovation activities benefit from the concentration of economic activities of similar and related firms in a cluster and facilitate knowledge spillovers and stimulate various forms of adaptation, learning and innovation (Skålholt and Thune 2013). The maritime cluster is of particular interest for the European Union as an area of potential economic valorisation connecting traditional sectors with science-based activities (Pinto et al. 2015a, b).

The mechanisms of knowledge production and exchange among the different agents that form a regional innovation system or a cluster are varied. They involve multiple processes or activities and different types and forms of knowledge flows and interactions among them. There is also an ample spectrum of innovation and knowledge exchange support services aimed at both encouraging knowledge transfer and reducing the barriers that all actors, but in particular SMEs, usually face in carrying out innovation (Pinto and Fernández-Esquinas 2013). The literature has amply stressed the importance of small business services, most notably the so called "real services" (Bellini 2003), namely support for business development, manufacturing and innovation processes, generally delivered directly with companies to stimulate knowledge transfer. Shapira et al. (2015) discuss the relevance of technology extension service (TES), defined as assistance provided directly to enterprises to foster technological modernization and improvement, with a focus on established SMEs.

Peripheral areas tend to present a less developed innovation support ecosystem, in particular because they are characterized by a large number of SMEs in less intensive technology sectors. SMEs in those areas may therefore find access to specialist knowledge provision problematic (Fernández-Esquinas et al. 2015). Given this 'knowledge intensive business services-poor' landscape, universities and public research organizations tend to play a stronger role as providers of specialist knowledge for regional companies (Pinto et al. 2015a, b).

Many factors influence the extent to which firms are able to benefit from the presence of an innovation support ecosystem. For instance, structural factors such as size and R&D expenditure affect the degree to which SMEs, and firms in general draw from external sources of knowledge. Sector specific dynamics also play a significant role in shaping the type and variety of knowledge interactions (Laursen and Salter 2006). Regarding typical barriers or constraints to the establishment of knowledge exchange interactions, communication barriers, appropriability problems, lack of absorptive capacity, and cultural differences, are the most frequently quoted in the literature (for a review see Perkmann et al. 2013).

As it is widely recognised in the literature, innovation is more difficult for SMEs than for larger firms: they usually have fewer resources, have less capacity to invest in R&D, and are in general more affected by uncertainty and innovation barriers (Bluhm and Schmidt 2008). The integration of SMEs into knowledge sharing networks and innovation systems, particularly at the regional level, constitutes a way to address these innovation difficulties (Teixeira et al. 2008). However, it is not often that SMEs truly engage in innovation networks, and in the cases in which they do their interaction tends to be mostly with business partners an much less with other knowledge providers in the network (i.e. universities, public research organizations and technology centres, public authorities and large firms). Whilst collaboration with business partners, such as customers and suppliers, is important to stimulate innovation in SMEs (Hassink 1997), knowledge exchange with other agents is also key as it allows SMEs to make use of all the potential sources of knowledge offered by their regional innovation systems environment (Zeng et al. 2010).

4.3 Methodological Notes

As indicated in the introduction, our analysis uses information collected from an online survey built with Qualtrics. The survey—provided in English, Portuguese and Spanish to facilitate understanding and encourage a high completion rate—was sent to potential users and/or providers of innovation or knowledge related exchange services operating in the maritime sector. These were 1743 entities located in France, the UK, Ireland, Spain, and Portugal (667 with valid e-mail contact).

The survey was active in the first half of 2014. The total number of responses received during this period was 102 from 491 entities which were able to receive the email with the invitation to complete the survey online. The response rate (20.7%) is acceptable in particular given the generality of the questions included in the survey and the heterogeneity in terms of type, sector, and characteristics of the entities forming the target population.

The survey was structured in four sections. The first section, inspired in the literature on university-industry interactions (Perkmann et al. 2013) included a set of general questions about the use (or lack thereof) of the following innovation or knowledge exchange and support services/schemes:

- Technical services and facilities (e.g. for certification, testing, prototyping, calibration)
- Business services and intelligence (marketing, access to markets, exporting)
- Innovation management advice (product/process innovation) and training
- Other training services related to innovation
- Incubation facilities and services (e.g. in science parks)
- R&D services
- Funding for co-operative R&D projects
- Services for inter-firm collaboration and networking (e.g. cluster associations)
- Provision of risk capital (venture capital, seed capital)
- Services/advice related to intellectual property protection
- Student placements or other type of mobility schemes between industry and universities/research organizations
- Other (please specify)

The category 'Other' was provided to allow respondents to add services or schemes not covered in the previous list.

The first section also included questions about potential reasons for not using these services, such as unawareness of their availability, high cost of the services, low quality and/or sophistication, lack of alignment with the organizations' needs, level of bureaucracy involved with their use, and the effects that the economic crisis has had on the organizations' investment capacity in this type of services.

The second section included questions in relation to the provision of the previously listed knowledge exchange services or schemes during the last 3 years. Organizations operating in the maritime sector that identified themselves as providers were enquired about:

- Their main clients and the percentage that they represent in the organization's portfolio (private firms, universities, technology centres and other research organizations, other public organizations, not for profit organizations, . . .);
- The location of their clients (local/regional, national, international);
- If they had detected any changes in the demand of these knowledge services;
- And in which ways they normally advertise their services (visits to clients, media, website, mailing lists, workshops. . .).

The third section of the survey focused on issues related to the use of specific knowledge services or schemes during the last 3 years. In particular, we asked organizations that identified themselves as users of these services:

- The importance of these services and activities for the organization (from 1 not important to 5 extremely important);
- The types of providers of the services they use (private firms, universities, technology centres and other research organizations, other public organizations, business angels,...);
- The location of these providers (local/regional, national, international);
- An assessment of the service (from 1 very poor to 5 very good) in terms of its quality, technical expertise of the staff delivering the service, its cost and accessibility, and its alignment with the organization's needs;
- And in which way their demand for these services had varied in the last 3 years.

The final section of the survey included a set of questions about the particular characteristics of the respondent organizations: their type, size, location, and the maritime sector to which they belong.[2]

4.4 Knowledge Exchange and Resilience of Innovation

4.4.1 Some Descriptive Results

The organizations that completed the survey belong to a high variety of sub-sectors within the maritime sector: sailing associations, sea biotechnology, health/wellbeing, maritime tourism, processing of sea salt, seaweed extracts services, fishing software, subsea and diving works, processing of fishing products, sails manufacturing, and wave energy sector. Most of the 102 organizations that completed the survey were located in different regions of Portugal (58%) followed by Ireland (13%), Spain (11%), France (11%), the UK (6%), and others (1%).

The respondents regarded different organization types (Fig. 4.1). Private enterprises (42.6%) were the most represented, followed by universities or research centres (22.1%), and not for profit organizations (11.8%). Regarding size, 40.6% had less than 10 employees, 18.9% 10–50 employees, 13.0% 50–250 employees, 15.9% 250–500 employees, and 11.6% more than 500 employees. 34% of the organizations did not used knowledge exchange activities or services used in the last 3 years. Many of the users of the listed services are also providers of some innovation support services themselves (see Table 4.1).

When prompted about the degree of importance of these activities for innovation, a few activities were more often reported by respondents as important or very

[2] The descriptive statistics report "Knowledge needs and innovation in the maritime economy" with interim data collection is available in the project website.

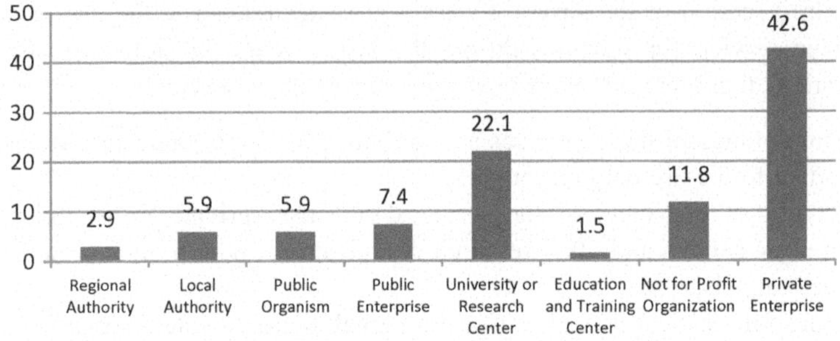

Fig. 4.1 Types of respondent organizations (in % of total answers). Source: Own elaboration

Table 4.1 Use and provision of knowledge exchange activities

		Does your organization provide any knowledge/exchange services?	
Does your organization use any knowledge/exchange services?		No	Yes
	No	33%	1%
	Yes	32%	33%

Source: Own elaboration

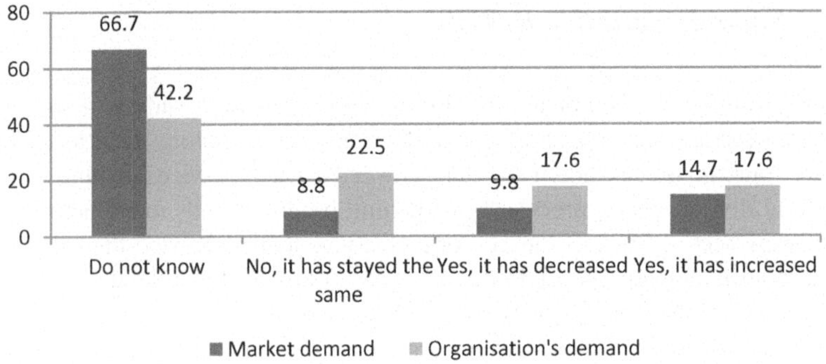

Fig. 4.2 Variations of market and organization's demands. Source: Own elaboration

important, namely funding for cooperative R&D projects and R&D services, followed by services for inter-firm collaboration and networking.

Two crucial variables for our study are the ones that look for the variations in market demand for innovation and knowledge services (In your experience, has the demand for these services varied in the last 3 years?) and the organization's demand (How has your demand for these services varied in the last 3 years?) (Fig. 4.2). The first element to retain is a high level of ignorance about this market (66.7% claim not to know what is happening in the market while 42.2% is unaware of the internal demand dynamics). Nonetheless, for those that are aware of what is happening, a clear majority considers that demand has stayed the same or even increased.

4.4.2 *Parametric and Non–Parametric Evidence*

Based in the KIMERAA survey we have selected specific variables to deepen the analysis using parametric and non-parametric techniques.[3] The goal was to get statistical evidence of the differences between groups of the organizations, namely, by size (18.6% have more than 250 workers), type (15.7% are academic versus remaining non-academic), and knowledge management (36.3% are organizations that provide or administer knowledge exchange services or schemes).

We present below the descriptive statistics for the selected variables (Table 4.2). Particularly relevant are the variables that try to detect the breadth of services used and provided by the organizations. These are count variables that sum if the organization uses/provides a particular type of service from the defined list.[4]

We tried to test if organizations with different size have the same utilization and provision of innovation and knowledge-based services. The breadth of services used and provided is presented in Fig. 4.3.

We rejected the hypotheses of equal means by different sizes, meaning that organization with different number of workers have provided and used a different breadth of mechanisms.[5] Both utilisation and provision grow with size but reach its maximum in medium-sized firms (50–250 workers) declining in bigger organizations.

Table 4.2 Descriptive statistics

	N	Minimum	Maximum	Mean	Deviation
USE—Variety of knowledge exchange activities or services used in the last 3 years	102	0.00	11.00	3.2941	3.28690
PROVISION—Variety of knowledge exchange activities or services provided in the last 3 years	102	0.00	11.00	1.5490	2.58950
CLIENT_EXPORT—Clients located internationally (%)	102	0.00	91.00	4.8824	16.98881
CLIENT_FIRM—Clients are MNEs and SMEs (%)	102	0.00	100.00	14.803	31.513
EVAL—Assessment of the quality of services used	102	0.00	30.00	11.5686	11.27646

Source: Own elaboration

[3]We used for this section the IBM SPSS Statistics 21.

[4]These two variables do not follow a normal distribution. The graphical intuition provided but the Q-Q plots and histograms is confirmed by the Kolmogorov-Smirnov test (1.869 and 3.759 compared to $n > 40$ and Sig $1\% = 0.25205$) (see histograms in Appendix).

[5]Looking for the homogeneity of variances, Levene test does not reject its null hypotheses of groups having homogeneous variances for the variable "utilisation". In this case ANOVA is valid (results in Appendix). But for "provision", the test rejects this H0 meaning that we need to use a non-parametric technique. We used Kruskal-Wallis that reinforced the findings (table test is also presented in Appendix).

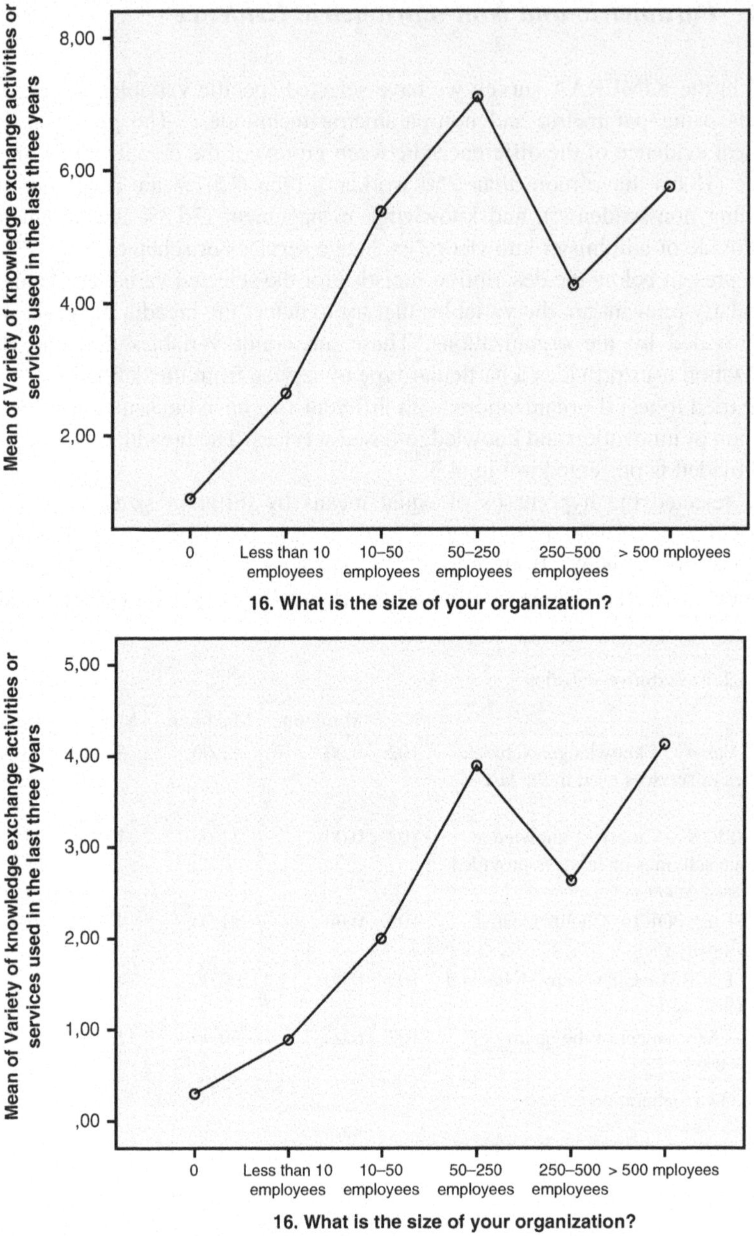

Fig. 4.3 Variety of utilisation and provision by size. Source: Own elaboration

In our sample, 36.3% of the organizations provided or administered knowledge exchange services or schemes (dummy variable 'KManag'). We used Chi-Square association tests to verify if entities that managed or administered these kinds of innovation services or schemes felt differently the variation in their utilisation and provision. We detected a significant association between these variables (cf. Appendix).

Using the non-parametric Mann-Whitney test we also have found that universities and other PROs are different of other types of organizations regarding their utilization and provision of knowledge services (cf. Appendix).

4.4.3 Econometric Evidence

In this section, we intend to consider a specific approach to resilience, by focusing the resilience of innovation. As explained before, in our understanding resilience of innovation operates at different levels, from the regional system to the innovation actors. The notion of resilience of innovation at system-level regards the capacity of a specific innovation system to deal with a negative disruption (external or systemic) and continue or improve its function. A strict and operational definition of innovation systems' resilience can be the capacity of the system to maintain innovative activities avoiding structural negative impacts of economic crises, namely those deriving from the contraction of regional product and the rise of unemployment. At micro-level, resilience of innovation regards the capacity of the firm or other relevant actor to continue or to improve innovative efforts despite external and internal shocks.

Based on these results the next step was to create a model that could explain the resilience of innovation in the organizations. For this purpose we considered as an operational definition, those organizations that experienced an increased or the equal demand of innovation and knowledge-based services during the last 3 years. A new binary variable was created with this objective. Then we transformed some the previously presented variables in more readable "dummies" for econometric analysis.[6]

The model used a Probit estimator. The results of three versions of the estimation are provided in the Table 4.3 below.

The model tries to assess the influence that internal (organizational related factors), external factors, and the innovation and knowledge services have in organization to the resilience of its innovation process. External factors refer to the context (cluster/environment in which the organizations operate), encapsulated by the market variation (MARKET_VARIATION), to detect the general situation, and the assessment of the quality of the services (EVAL) as a proxy of the sophistication of the existing supply. Internal factors relate to organizational capabilities, and are represented by types of clients, exports, size, academic profile, knowledge management. We also pay attention to the influence of the breadth of utilization (USE) and provision (PROVISION) of innovation and knowledge services to the resilience of the process.

[6]"SIZE_BIG" is a binary variable that assumes the value 1 if the organization has 250 or more workers. "UNIV_PROS" assumes 1 if organization is a university or other PRO. "MARKET_VARIATION" is a dummy that assumes value 1 if organizations believe that their market experienced an increased or at least an equal demand of innovation and knowledge-based services during the last 3 years.

Table 4.3 Probit model

Variable	Model 1 Global	Model 2 Eliminated non-significant variables	Model 3 Market variation effects eliminated
C	−2.573697***	−2.542377***	−1.908418***
External factors			
MARKET_VARIATION	3.515976***	3.335156***	–
EVAL	0.083005***	0.080570***	0.062051***
Internal factors			
CLIENT_FIRM	−0.003265	–	–
CLIENT_EXPORT	0.054807**	0.048929*	0.026621
SIZE_BIG	1.503512**	1.855549***	1.272323***
UNIV_PROS	0.681733	–	–
KMANAG	−1.910828**	−1.681732**	0.723818
Innovation and knowledge services			
USE	0.252214**	0.254233**	0.192370**
PROVISION	−0.565364***	−0.540563***	−0.173323*

Source: Own elaboration
61 Obs with Dep = 0; 41 Obs with Dep = 1
*Significant at 0.1, **significant at 0.05, ***significant at 0.01

Our exploratory results suggest that in terms of internal factors, the resilience of innovation is positively influenced by size and by exports. The fact that organizations manage innovation or knowledge exchange schemes is statistically significant but with a negative impact.

The influence of the breadth of innovation and knowledge services shows an interesting effect. While organizations that use a larger number of types of services have more resilient innovation processes, the provision of a larger number of types of services has a negative impact on their demand.

4.5 Conclusion

The maritime cluster is of particular interest for the European Union as an area of potential economic valorisation connecting traditional sectors with science-based activities. In this chapter, drawing upon empirical date on the knowledge provision and needs of maritime cluster innovative organizations in the European Atlantic Area, we have provided econometric evidence of the main internal and external factors that influence the resilience of innovation at the organizational level. We defined 'resilience of innovation' as the multi-level capacity of the innovation process to maintain or accelerate its functions when facing an internal disruption or an external shock.

In terms of internal factors, the resilience of innovation at organizational level is positively influenced by size—this confirms ideas found in the literature i.e. small firms have more difficulties in being innovative, and by exports—suppliers' interaction favour knowledge exchange and creation, also supported by existing literature on learning and supply chains. The fact that the dummy variable that regards to organizations that manage innovation or knowledge exchange schemes is statistically significant but with a negative impact could be due to opportunity cost of dedicating time to managing: time and managerial attention are scarce resources.

Regarding the influence of the breadth of innovation and knowledge services, our results indicate that organizations that use a larger number of types of services have more resilient innovation processes, this could be because these organizations are innovative resilient. They use this variety to generate knowledge.

The provision of a larger number of types of services has a negative impact. Perhaps due to cost of keeping multiple sources at a time of economic crisis when less variety is demanded in general, by most organizations (even if those among them that are resilient demand it).

Our results confirm the relevance of innovation and knowledge service provision. Even—or particularly more so—in times of economic crisis, increasing access to these services is key considering the importance of these services for the resilience of innovation. This suggests a need to better communicate to firms, particularly small firms, the availability of such services and improve the alignment and accessibility of these services by the smallest firms.

Regional policy actors may consider expanding and better connecting the network of service provision with the needs of firms and providing additional funding and other incentives to encourage the use of these services, for instance by using mechanisms such as innovation vouchers.

Some organizations have witnessed an increase of demand for their services. These activities could be further promoted and enabled, through public programmes for knowledge transfer and technology extension infrastructure, including efforts to build up long-term capacity for the provision of services that are customized to the needs of client firms, yet adaptive and flexible, as well as the development of good knowledge exchange networks between clients, service providers and other resources.

In some peripheral areas, local and regional universities tend to partly compensate for the relative lack of other private knowledge intensive service and venture capital providers regionally and their role in enabling innovation of local enterprises should be acknowledged and supported by public policy. Firms would also benefit from the reduction of red tape and unnecessary bureaucracy associated with innovation support. Finally, firms, particularly micro enterprises, may lack the absorptive capacity to benefit from the presence of innovation support infrastructure, particularly research-intensive activities from universities. A clearer development pathway may be promoted that builds up the competences of these firms through low-level services and eventually allows an upgrade to more sophisticated, research-intensive activities.

Acknowledgements This chapter was stimulated by KIMERAA—Knowledge transfer to Improve Marine Economy in Regions from the Atlantic Area, project developed between 2011 and 2014, co-financed by the European cooperation program INTERREG Atlantic Area through the ERDF—European Regional Development Fund. The help of project partners in the data collection is gratefully acknowledged. Carla Nogueira benefits from the financial support from FCT—Portuguese Science and Technology Foundation (SFRH/BD/117398/2016). Hugo Pinto also acknowledges the financial support from FCT (SFRH/BPD/84038/2012).

Appendix

Histogram of Variety of Uses and Provisions

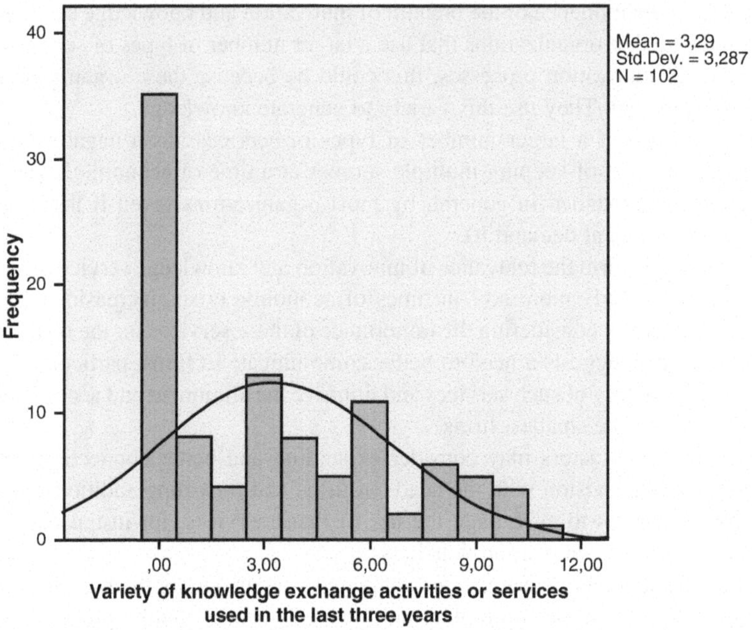

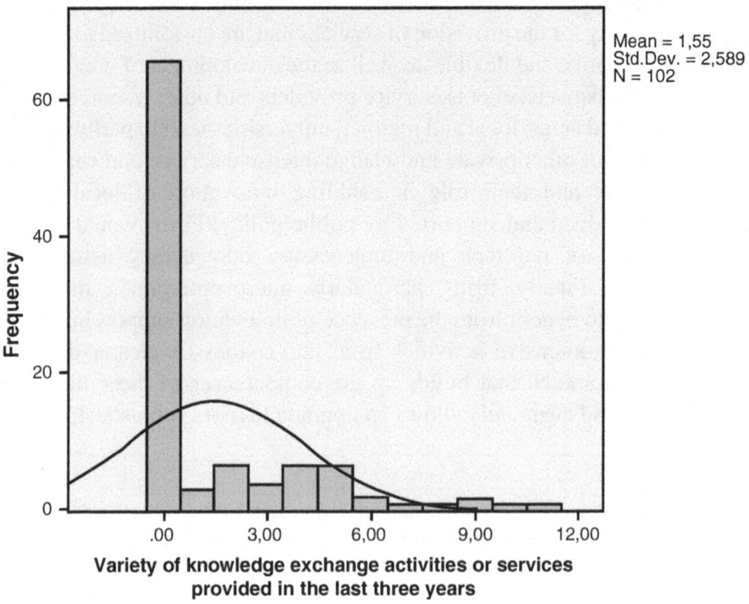

Source: Own elaboration

Anova

		Sum of squares	df	Mean square	F	Sig.
Variety of knowledge exchange activities or services used in the last 3 years	Between groups	423.221	5	84.644	12.165	.000
	Within groups	667.955	96	6.958		
	Total	1091.176	101			
Variety of knowledge exchange activities or services provided in the last 3 years	Between groups	181.297	5	36.259	7.019	.000
	Within groups	495.958	96	5.166		
	Total	677.255	101			

Source: Own elaboration

Kruskal-Wallis Test

	Size of organization	N	Mean rank
Variety of knowledge exchange activities or services used in the last 3 years	0	33	29.26
	Less than 10 employees	28	48.00
	10–50 employees	13	70.08
	50–250 employees	9	84.61
	250–500 employees	11	62.09
	> 500 employees	8	73.50
	Total	102	
Variety of knowledge exchange activities or services provided in the last 3 years	0	33	37.80
	Less than 10 employees	28	47.66
	10–50 employees	13	57.15
	50–250 employees	9	79.44
	250–500 employees	11	59.23
	> 500 employees	8	70.19
	Total	102	

Source: Own elaboration

Kruskal Wallis test	Variety of knowledge exchange activities or services used in the last 3 years	Variety of knowledge exchange activities or services provided in the last 3 years
Chi-Square	43.218	27.451
Df	5	5
Asymp. Sig.	0.000	0.000

Source: Own elaboration

Notes: Kruskal Wallis test, Grouping variable: what is the size of your organization?

Mann–Whitney Test

	Universities and PROs	N	Mean rank	Sum of ranks
Variety of knowledge exchange activities or services used in the last 3 years	Other	86	47.62	4095.50
	University or PRO	16	72.34	1157.50
	Total	102		
Variety of knowledge exchange activities or services provided in the last 3 years	Other	86	46.28	3980.00
	University or PRO	16	79.56	1273.00
	Total	102		

Source: Own elaboration

	Variety of knowledge exchange activities or services used in the last 3 years	Variety of knowledge exchange activities or services provided in the last 3 years
Mann-Whitney U	354.500	239.000
Wilcoxon W	4095.500	3980.000
Z	−3.140	−4.842
Asymp. Sig. (2-tailed)	0.002	0.000

Source: Own elaboration
Note: Grouping variable: Universities and PROs

Tests for Independence

Association of between "Does your organization provide or administer any knowledge exchange services or schemes?: * Variety of knowledge exchange activities or services used in the last 3 years".

Chi-square tests			
	Value	df	Asymp. Sig. (2-sided)
Pearson Chi-square	103.831[a]	33	0.000
Likelihood ratio	112.200	33	0.000
Linear-by-linear association	5.112	1	0.024
N of valid cases	102		

Source: Own elaboration
[a]45 cells (93.8%) have expected count less than 5. The minimum expected count is 0.04

Symmetric measures			
		Value	Approx. Sig.
Nominal by nominal	Contingency coefficient	0.710	0.000
N of valid cases		102	

Source: Own elaboration

Association between "Does your organization provide or administer any knowledge exchange services or schemes? * Variety of knowledge exchange activities or services provided in the last 3 years".

Chi-square tests			
	Value	df	Asymp. Sig. (2-sided)
Pearson Chi-square	97.740[a]	33	0.000
Likelihood ratio	123.252	33	0.000
Linear-by-linear association	1.394	1	0.238
N of valid cases	102		

Source: Own elaboration

[a]45 cells (93.8%) have expected count less than 5. The minimum expected count is 0.04

Symmetric measures			
		Value	Approx. Sig.
Nominal by nominal	Contingency coefficient	0.700	0.000
N of valid cases		102	

Source: Own elaboration

Predictive Capacity of Probit Model

Model in E-Views: resilience c client_firm clients_export size_big univ_pros kmanag eval use provision market_variation.

Global Model

Mean dependent var	0.401961	S.D. dependent var	0.492715
S.E. of regression	0.309832	Akaike info criterion	0.711688
Sum squared resid	8.831618	Schwarz criterion	0.969038
Log likelihood	−26.29607	Hannan-Quinn criter.	0.815898
Restr. log likelihood	−68.72747	Avg. log likelihood	−0.257805
LR statistic (9 df)	84.86279	McFadden R-squared	0.617386
Probability(LR stat)	1.74E-14		
Obs with Dep = 0	61	Total obs	102
Obs with Dep = 1	41		

Source: Own elaboration

Global model prediction evaluation (success cutoff C = 0.5)						
	Estimated equation			Constant probability		
	Dep = 0	Dep = 1	Total	Dep = 0	Dep = 1	Total
P(Dep = 1) < =C	55	7	62	61	41	102
P(Dep = 1) > C	6	34	40	0	0	0
Total	61	41	102	61	41	102
Correct	55	34	89	61	0	61
% Correct	90.16	82.93	87.25	100.00	0.00	59.80
% Incorrect	9.84	17.07	12.75	0.00	100.00	40.20
Total Gain[a]	−9.84	82.93	27.45			
Percent Gain[b]	NA	82.93	68.29			
	Estimated equation			Constant probability		
	Dep = 0	Dep = 1	Total	Dep = 0	Dep = 1	Total
E(# of Dep = 0)	52.69	8.58	61.27	36.48	24.52	61.00
E(# of Dep = 1)	8.31	32.42	40.73	24.52	16.48	41.00
Total	61.00	41.00	102.00	61.00	41.00	102.00
Correct	52.69	32.42	85.11	36.48	16.48	52.96
% Correct	86.38	79.07	83.44	59.80	40.20	51.92
% Incorrect	13.62	20.93	16.56	40.20	59.80	48.08
Total Gain[a]	26.57	38.87	31.52			
Percent Gain[b]	66.11	65.00	65.56			

Source: Own elaboration

[a]Change in "% Correct" from default (constant probability) specification

[b]Percent of incorrect (default) prediction corrected by equation

References

Alexander, D. E. (2013). Resilience and disaster risk reduction: An etymological journey. *Natural Hazards and Earth System Sciences, 13*(11), 2707–2716. https://doi.org/10.5194/nhess-13-2707-2013

Archibugi, D., Filippetti, A., & Frenz, M. (2013). Economic crisis and innovation: Is destruction prevailing over accumulation? *Research Policy, 42*(2), 303–314. https://doi.org/10.1016/j.respol.2012.07.002

Asheim, B. T., Coenen, L., Moodysson, J. & Vang, J. (2005). *Regional innovation system policy: A knowledge-based approach*. Papers in Innovation Studies 2005/13, Lund University, CIRCLE – Center for Innovation, Research and Competences in the Learning Economy.

Asheim, B. T., Smith, H. L., & Oughton, C. (2011). Regional innovation systems: Theory, empirics and policy. *Regional Studies, 45*(7), 875–891. https://doi.org/10.1080/00343404.2011.596701

Bellini, N. (2003). *Business support services: Marketing and the practice of regional innovation policy*. Cork: Oak Tree Press.

Bluhm, K., & Schmidt, R. (2008). *Change in SMEs – The new European capitalism*. Basingstoke: Palgrave Macmillan. https://doi.org/10.1057/9780230227781

Boschma, R. (2005). Proximity and innovation: A critical assessment. *Regional Studies, 39*(1), 61–74. https://doi.org/10.1080/0034340052000320887

Boschma, R. (2015). Towards an evolutionary perspective on regional resilience. *Regional Studies, 49*(5), 733–751. https://doi.org/10.1080/00343404.2014.959481

Boschma, R., & Martin, R. (2010). *The aims and scope of evolutionary economic geography*. Papers in Evolutionary Economic Geography (PEEG). Retrieved from http://ideas.repec.org/p/egu/wpaper/1001.html

Cooke, P. (2013). *Complex adaptive innovation systems: Relatedness and transversality in the evolving region*. Oxford: Routledge.

Cooke, P., & Leydesdorff, L. (2006). Regional development in the knowledge-based economy: The construction of advantage. *Journal of Technology Transfer, 31*, 5–15.

Davoudi, S., Shaw, K., Haider, L. J., Quinlan, A. E., Peterson, G. D., Wilkinson, C., et al. (2012). Resilience: A bridging concept or a dead end? "reframing" resilience: Challenges for planning theory and practice interacting traps: Resilience assessment of a pasture management system in Northern Afghanistan urban resilience: What does it mean in Planni. *Planning Theory & Practice, 13*(2), 299–333.

Doloreux, D. (2002). What we should know about regional systems of innovation. *Technology in Society, 24*(3), 243–263.

Fernández-Esquinas, M., Pinto, H., Yruela, M. P., & Pereira, T. S. (2015). Tracing the flows of knowledge transfer: Latent dimensions and determinants of university–industry interactions in peripheral innovation systems. *Technological Forecasting and Social Change*. https://doi.org/10.1016/j.techfore.2015.07.013

Frenken, K., Van Oort, F., & Verburg, T. (2007). Related variety, unrelated variety and regional economic growth. *Regional Studies, 41*(5), 685–697. https://doi.org/10.1080/00343400601120296

Frenz, M., & Prevezer, M. (2012). What can CIS data tell us about technological regimes and persistence of innovation? *Industry and Innovation, 19*(4), 285–306. https://doi.org/10.1080/13662716.2012.694676

Gertler, M. S. (2003). Tacit knowledge and the economic geography of context, or the undefinable tacitness of being (there). *Journal of Economic Geography, 3*(1), 75–99.

Hamdouch, A., & Depret, M. (2012). *Mondialisation et résilience des territoires*. Québec: Presses de l'Université du Québec.

Hassink, R. (1997). Technology and transfer infrastructures: Some lessons from experiences in Europe, the US and Japan. *European Planning Studies, 5*, 351–370.

Howells, J. (2005). Innovation and regional economic development: A matter of perspective? *Research Policy, 34*(8), 1220–1234. https://doi.org/10.1016/j.respol.2005.03.014

Laursen, K., & Salter, A. (2006). Open for innovation: The role of openness in explaining innovation performance among U.K. manufacturing firms. *Strategic Management Journal, 27* (2), 131–150. https://doi.org/10.1002/smj.507

Makkonen, T. (2013). Government science and technology budgets in times of crisis. *Research Policy, 42*(3), 817–822. https://doi.org/10.1016/j.respol.2012.10.002

Martin, R., & Sunley, P. (2015). On the notion of regional economic resilience: Conceptualization and explanation. *Journal of Economic Geography, 15*, 1–42. https://doi.org/10.1093/jeg/lbu015

Paunov, C. (2012). The global crisis and firms' investments in innovation. *Research Policy, 41*(1), 24–35. https://doi.org/10.1016/j.respol.2011.07.007

Perkmann, M., Tartari, V., McKelvey, M., Autio, E., Broström, A., D'Este, P., et al. (2013). Academic engagement and commercialisation: A review of the literature on university–industry relations. *Research Policy, 42*(2), 423–442. https://doi.org/10.1016/j.respol.2012.09.007

Pinto, H., & Fernández-Esquinas, M. (2013). Exploring knowledge-transfer dynamics in a South European region: Breadth, intensity and informality of university-industry interactions in Andalusia: Knowledge commercialization and valorization in regional economic development. In T. Baycan (Ed.), *Knowledge commercialization and valorization in regional economic development*. Edward Elgar. Retrieved from http://www.elgaronline.com/abstract/9781781004067.00018.xml

Pinto, H., & Pereira, T. S. (2014). Resiliência dos sistemas de inovação face à turbulência económica. *Oficina Do CES, 418* Retrieved from http://www.ces.uc.pt/publicacoes/oficina/index.php?id=11157

Pinto, H., Cruz, A. R., & Combe, C. (2015a). Cooperation and the emergence of maritime clusters in the Atlantic: Analysis and implications of innovation and human capital for blue growth. *Marine Policy, 57*, 167–177. https://doi.org/10.1016/j.marpol.2015.03.029
Pinto, H., Fernandez-Esquinas, M., & Uyarra, E. (2015b). Universities and knowledge-intensive business services (KIBS) as sources of knowledge for innovative firms in peripheral regions. *Regional Studies, 49*(11), 1873–1891. https://doi.org/10.1080/00343404.2013.857396
Rodríguez-Pose, A. (2013). Do institutions matter for regional development? *Regional Studies, 47* (7), 1034–1047. https://doi.org/10.1080/00343404.2012.748978
Shapira, P., Uyarra, E., Youtie, J., Cox, D., Gok, A., Rogers, J., et al. (2015). *Institutions for Technology Diffusion*, USA. Inter-American Development Bank, Competitiveness and Innovation Division. Report No. IDB-TN-832.
Simmie, J. (2014). Regional economic resilience: A Schumpeterian perspective. *Raumforschung und Raumordnung, 72*(2), 103–116. https://doi.org/10.1007/s13147-014-0274-y
Simmie, J., & Martin, R. L. (2010). The economic resilience of regions: Towards an evolutionary approach. *Cambridge Journal of Regions Economy and Society, 3*, 27–43.
Skålholt, A., & Thune, T. (2013). Coping with economic crises—The role of clusters. *European Planning Studies, 22*(10), 1993–2010. https://doi.org/10.1080/09654313.2013.813909
Teixeira, A. C., Santos, P., & Oliveira Brochado, A. (2008). International R&D cooperation between low-tech SMEs: The role of cultural and geographical proximity. *European Planning Studies, 16*(6), 785–810. https://doi.org/10.1080/09654310802079411
Tödtling, F., & Trippl, M. (2005). One size fits all? *Research Policy, 34*(8), 1203–1219. https://doi.org/10.1016/j.respol.2005.01.018
Torre, A., & Rallet, A. (2005). Proximity and localization. *Regional Studies, 39*(1), 47–59. https://doi.org/10.1080/0034340052000320842
Uyarra, E., & Flanagan, K. (2012). Reframing regional innovation systems: Evolution, complexity and public policy. In P. Cooke (Ed.), *Re-framing regional development* (pp. 146–163). London: Routledge.
Zeng, S. X., Xie, X. M., & Tam, C. M. (2010). Relationship between cooperation networks and innovation performance of SMEs. *Technovation, 30*(3), 181–194. https://doi.org/10.1016/j.technovation.2009.08.003

Chapter 5
Innovation, Regions and Employment Resilience in Sweden

Charlie Karlsson and Philippe Rouchy

5.1 Introduction

The concept of regional resilience draws currently a lot of attention in the context of the ability of territories to recover from economic crisis. Currently, one sees theoretical and empirical researches on resilience reaching no consensus on a privileged line of inquiry. It is original form, resilience carries its macroeconomic meaning whereby national economies recover from recessions and other economic shocks (see Bristow 2010; Cellini and Torrisi 2014; Christopherson et al. 2010; Fingleton et al. 2012, 2015; Hassink 2010; Hill et al. 2008; Martin 2012; Martin and Sunley 2015; Ormerod 2008, 2010; Simmie and Martin 2010). Resilience is calling attention to those moments of after-shock whereby markets responds to external disturbances through their return to equilibrium. In economic geography, Martin and Sunley (2015) and Martin (2012) have argued that one may better understand economic recession by gainfully complementing it with specificities of regional cyclical sensitivities. In using resilience, regional economics should be able to address shock by showing time lag in their reaction to disturbing causes, which took place earlier. In line with evolutionary economics, regional resilience captures the ability of regions to reconfigure their socio-economic structure over time (Christopherson et al. 2010; Simmie and Martin 2010; Cooke et al. 2011).

C. Karlsson (✉)
Jönköping International Business School, Jönköping, Sweden
e-mail: Charlie.Karlsson@ju.se

P. Rouchy
Department of Industrial Economics, Blekinge Institute of Technology, Karlskrona, Sweden
e-mail: Philippe.Rouchy@bth.se

H. Pinto et al. (eds.), *Resilience and Regional Dynamics*, Advances in Spatial Science, https://doi.org/10.1007/978-3-319-95135-5_5

Considering the effects of regional adaptation to socio-economic reconfiguration leaves a lot of space for reflection. Many researchers (Martin 2010; Boschma 2015) argue that long-term adaptive capacities of regions are an open research agenda. In line with their argument, we propose a historically based assessment of resilience based on six selected Swedish regions. We temper our conceptual theorizing by showing some basic empirical insights on regional behaviour aiming at furthering the reflection on regions' resilience and adaptations to change. First, we conceptualize regional resilience as regions' abilities to adapt to continuous changes over time through regional labour characteristics (regional net employment, accessibility defined as commuting surplus/deficit and labour dynamics private/public). In this sense, we ground directly the ability of regions to reconfigure their socio-economic settings into aspects of labour economics. Second, we propose to use this framework as a basis for some preliminary empirics based on six most innovative Swedish regions. Those six regions show a spread distribution of innovation from "highly innovative" to "followers in innovation" (Table 3.2). In taking up the question of how labour affects the ability of regions to reconfigure themselves, we question the relevance of economic shocks to anchor the analysis of regional resilience. We think that regional labour market captures the most relevant aspects of regional resilience behaviour. Labour market aspects of resilience has received little attention in the literature so far (Diodato and Weterings 2014; Fingleton et al. 2012) mainly because labour economics has been at the periphery of some of the most recent dimensions of regional economics, namely the dynamics of industries, networks and institutions. In addition, we make a point in this study to consider the development of regions overtime as a focal point to capture our concept of regional resilience. For that matter, the study covers a period of 10 years between 2004 and 2014. This period displays regional labour market characteristics showing dependency on pre-existing industrial and other regional institutions but also localized changes. The timeframe provides a mean to assess the relative importance of disturbance—and if such disturbances are identifiable as economic shock. Our view of disturbance is in line with Ormerod's (2010) long term historical findings (between 1871 and 2007) showing that most recessions in western economies last for 1 year. Consequently, descriptive regional characteristics of the labour market aim at defining more relevant aspects of regional resilience than currently in use.

This chapter is organized as follows: In Sect. 2, we discuss the theoretical treatment of resilience in the literature. We propose a time sensitive approach of regional resilience in which labour market defines key aspects of socio-economic condition of resilience. In Sect. 3, we present basic characteristics of the Swedish innovative regions according to NUTS3 and our selection of six of them. In Sect. 4, we are gathering some preliminary empirics on those six Swedish regions to show how regional resilience can be defined through labour market's performance, dynamics and accessibility. In Sect. 5, we conclude on policy implications and suggestions for further studies.

5.2 Toward Evolutionary Based Notions of Regional Resilience

Resilience has been of great use by economists and diverse social scientists to talk about recovery from economic shock and responsiveness of individuals and organization to sudden changes. The definition of resilience refers to the ability of a system or entity to recover its original form or regain its position after disturbance or disruption. In the regional economic literature (Foster 2007; Hill et al. 2008), the focus has been on socio-economic system recovering from disruption or shock. Despite the relative meaning of economic shock, there is a majority of the literature observing diverse degrees of regional absorption. Economic geographers have covered issues of resilience in regional case study (Treado 2010), comparative analysis of regions (Swanstrom et al. 2009; Simmie and Martin 2010; Wolfe 2010; Hill et al. 2012) and system approaches (Diodato and Weterings 2014; Fingleton et al. 2012; Martin 2012). There are different reformulations of the resilience concept notably its ecological version (Reggiani et al. 2002; Swanstrom et al. 2009; Zolli and Healy 2012). It is a reformulation of the neoclassic concept of equilibrium whereby a region is reaching a new equilibrium state after facing external shock. Many scholars (Christopherson et al. 2010; Clark et al. 2010; Pike et al. 2010; Simmes and Martin 2010; Cooke et al. 2011; Boschma 2015) have preferred an evolutionary approach. It distinguishes itself with the view that resilience is a long-term capacity of a region to adapt and reconfigure its industrial, technological and institutional structures given the ever-changing condition of the economy. The understanding of change is more in line with change in business cycles (rather than external shock). By contrast, Martin (2012) is proposing to understand resilience as a structural re-organisation of the industrial makeup of a region. For him, it is an adaptive process of anticipation and reaction to minimize the impact of shock that could have a destabilizing effect on the regional economy. Even if the theoretical elements of "adaptation" have been introduced, Martin is dealing with resistance to shock. He made that point explicit in distinguishing 4 dimensions of regional resilience: (i) regional resistance to disturbances and disruptions, (ii) speed and extend of recovery, (iii) a structural re-orientation of the region output, jobs and incomes and, (iv) the extent to which the region has renewed its economy to resume its growing path. His focus is on regional "adaptive resilience" as "the capacity of a regional economy to reconfigure, that is to adapt, its structure (firms, industries, technologies and institutions) to maintain an acceptable growth, employment and wealth". Such adaptability will depend on (i) the rate of entrepreneurship and new firm formation in the region (Andersson and Koster 2011), (ii) the innovativeness of existing firms and their ability and willingness to shift into new sectors and product lines, (iii) access to finance for investment, (iv) the diversity of the region's economic structure, and (v) the availability of labour of the right skills, and similar factors (Martin 2012: 10-1).

A more straightforward evolutionary interpretation of regional resilience is given by Simmie and Martin (2010) which consider the regional economic system to be resilient if it is considered an ongoing process rather than a recovery to some equilibrium state. Here is clearly emphasized the historical development of resilience over time to changing condition. It seems that resilience is closer to the traditional definition of regional innovation whereby regions cope with structural change by their ability to create new growth paths and challenge stagnations and decline by emphasizing other economic sectors (Saviotti 1996).

Other recent approach of regional resilience embraces an evolutionary perspective adding related varieties of regions to define the content of resilience. Boschma (Boschma and Lambooy 1999; Boschma 2015) defines resilience as a regional capacity with a long-term adaptability whereby history defined as a regional path dependency affects economic renewal but also helps overcome negative lock-ins. Boschma's treatment of regional resilience recognizes the role of history not necessarily as a negative constraining aspect of regional renewal. He proposes his version of adaptive resilience as a matrix of related industrial variety, which needs to be activated to secure regional resilience. For that matter, Boschma considers regional resilience to integrate three elements of renewal. Those are:

1. Techno-industrial variety—it deals with the problem of single-industry-regions in comparison with multi-industries regions and their ability to recombine themselves to generate new growth avenues (Neffke et al. 2011a, b; Essletzbichler 2015; Boschma et al. 2013; Neffke et al. 2014).
2. Knowledge networks whereby people in regions combine different sources of knowledge to create new knowledge. Regional proximity plays an essential role in human capital for establishing networks ties and decreasing costs and risks (Boschma and Frenken 2010; Balland 2012a, b). In this view, knowledge networks are one component of regional resilience.
3. Institutions are closely related to techno-industrial variety and networks. Particularly, institutions should help to adapt to change. It is reflected in the birth of new institutions for regional development. Boschma underlines the fact that institutions have been more carefully reconsidered lately due to their complementarity with other forms of growing industrial factors (Amable 2000; Hollingsworth 2000; Hall and Soskice 2001; Grillitsch 2014). Institutions are also considered source of adaptation and recombination with existing institutions (Ebbinghaus 2009; Strambach 2010; Strambach and Klement 2012; special issue on Zeitschrift fur Wirtschaftsgeographie 2013).

In such overreaching research programs on regional resilience, there are spaces for investigating human resources further. In the coming section, we will pick up three issues sensitive to regional resilience:

1. Regional resilience may display the ability of region to absorb economic shock but not only. For that matter, it is important to distinguish between regional resistance/recovery to external shock and regional ability to adapt to business cycles and develop new growth paths.
2. The second point is the role of historical and regional development. Generally, the role of history has been a problem of regional adaptability due to negative path dependencies inherited from the past. In other words, discrepancies exist

between traditional industrial path and adaptation to new industrial conditions. In this view, the focal point is the adjustment to new conditions (Magnusson and Ottosson 2009; Henning et al. 2013). In fact, this view is partly based on long term industrial change taking place during the 1980s (Markusen 1985; Doussard and Schrock 2015). In contrast, the 2007–2009 crisis has been dealt as a short-term contraction in the economy (Ormerod 2010; Graddy-Reed and Feldman 2015). The outcome of the evolutionary view on short term cyclical crisis in a regional context, is to take skills, resources, technologies and institutions as dynamic means of adaptation to new economic conditions (Andersson and Koster 2011).

3. The third point is related to the two previous ones: regional resilience is the result of a new set up in economic answer to shock, i.e. labour adaptation and institutional renewal. These dimensions were scarcely approached 10–20 years ago. Regional studies have investigated key variables in this area, namely the role of labour force and employment (Fingleton et al. 2012). One needs a complex and multi-layered definition of regional resilience, looking at meso-processes such as the role of the labour market in regional development. Labour markets do not simply play the role of an adjustment variable during economic shock (in this view, labour market is understood exclusively as a decline in private employment and public cutbacks).[1] There is room for addressing regional resilience in relation either to regional employment dynamics or dependence on the public-sector employment (Bristow 2010; Hassink 2010; Pike et al. 2010; Davies 2011), or to people's ability to answer trade-off between their living and working conditions. As such, a more dynamic approach of labour market embraces not only economic shock but also life-style choices (Graddy-Reed and Feldman 2015).

In the following Table 5.1, we synthesize three different perspectives in regional economics dealing with resilience.

The territorially embedded resilience was originally studied by the theorists of RIS (regional innovation system) (Asheim and Isaksen 1996, 2001; Cooke et al. 1997; Meeus et al. 1999: 9; Wiig 1996). One of the key concerns for regional innovation system is to consider political issues of regional development. Notably governments are concerned with the harmonization of regional disparities (rather than economic shock per se, sees European Commission, COM 2014). From that point of view, the regional level of analysis started to be considered as a complement of national innovation policies. Underlying the regional innovation system lays a concern for overcoming economic troubles but also to manage restructuration and to launch innovation policies. For that matter, some researchers (Karlsson and Olsson 2000; Andersson and Karlsson 2004) considered RIS to complement the concept of functional regions, since they share mechanisms of renewal. For example, RIS and

[1]Across, the 1980s and 1990s, economists have been amongst the first to criticize the idea that labour flexibility was meant to reflect unemployment in times of regional restructuring.

Table 5.1 Three areas of resilience in regional economics

Type of resilience	Definition of resilience	Economic level of analysis	Resources
RIS and functional regions	The ability of a territory to avoid lock-in situations (Asheim and Isaksen 2002; Andersson and Karlsson 2004)	Innovation as breaking path dependency and changing technological trajectory	Finding locally relevant knowledge institutions Geographical (commuting patterns), social (networks) and cultural proximity
Adaptive resilience–shock theory	The ability of a system to undergo anticipatory or reactionary reorganization to minimize impact of destabilizing shock. (Martin 2012; Boschma 2015)	Regional structure (firms, industries, technology, institutions) to maintain an acceptable growth (output, employment)	Schumpetarian creative destruction: disturbance, disruption, recession of firms, industries, technologies and institutions Opportunity of development of new sectors and adaptive capability of the regional economy depends on region pre-existing economy (path dependency, Martin 2010, Andersson and Koster 2011)
Labour market resilience	The ability of the labour markets to weather economic downturns with limited social costs. (Diodato and Weterings 2014; OECD 2012a, b, c)	Worker welfare–the ability to find a job after unemployment	The ability to find job after unemployment depends on inter-sectoral and interregional labour mobility Tax benefit systems on labour cost Coordination of wage bargaining institutions

Sources: Asheim and Isaksen 2002; Andersson and Karlsson 2004; Martin 2010, 2012; Boschma 2015; Diodato and Weterings 2014; OECD 2012a, b, c

functional regions' approaches share concerns regarding high intensity of economic interaction (Johansson 1998), the accessibility of municipalities to relevant economic networks and the networks of infrastructure (Johansson 1992, 1993). We want to emphasize the idea that regional "system" or "function" contain already key principles of resilience. For example, Almeida and Kogut (1999) show that labour market plays an adaptive role to change in maintaining flows of knowledge within regional labour networks. Not only it suggests that labour markets are not only key to foster regional change but able to generate their own answers to change. For example, regional economists have shown that commuting patterns can be regarded as an appropriate method to assess regional borders and interaction. A region displays resilience when a territory shows the capacity of its labour market to take advantage of learning processes though interaction within industrial clusters and benefiting from institutional support. In this chapter, the first dimension of regional resilience we explore is labour accessibility. Labour accessibility is defined as commuting pattern showing an organizational continuity between actors and flows of goods and services.

The other approaches of adaptive resilience are clearly related to a treatment of economic shocks (Fingleton et al. 2012; Martin 2012). The theoretical treatment of the question is essential concerned with the evolution of the long-run regional disparities. The concern seeks to know to what extend the negative effect of shocks affecting national growth (recessions, financial crises, political upheavals) can be observed at the regional level. In this perspective, the notion of resilience is an attempt to capture the reaction of regional economies (the meso-level) to major recessionary shocks (the macro-level). To a certain extent, it interrogates the possibilities of regions to react from a downturn and its ability to transform it into a rapid growth. This issue focuses on the regions' ability to create significant growth higher than its pre-shock rate. Traditionally, regional output growth uses similar indicators than national growth, i.e. production output, firm formation, new sector formation, employment rate, new sectoral productivity and science and technology innovation as well as indicators of institutional reforms (Caballero and Hammour 1994; Gali and Hammour 1993; Andersson and Koster 2011). In this chapter, we focus on regional growth and labour market characteristics. More specifically, we survey descriptive data on labour market efficiency defined as the share of public employment in total regional employment to assess the region's ability to grow.

The third aspect in the literature is the labour market resilience. The basic assumption is to consider a coupling between the recovery of firms in a region and access to the labour market. As reviewed before, one of the reasons for focusing on labour effect is its ability to adapt (through commuting pattern) to new situations. In this view, regional resilience defines the ability of a region to absorb in its job market the work force after an economic downturn (it may go from regional regeneration to reallocation of the workforce to other regions). It is not exaggerated to say that regional resilience is considering elements of regional revival (reviewed in Table 5.1, third section on labour market resilience). McCann and Ortega-Argilés (2013) have synthesized the regional resilience by considering embeddedness, skill-relatedness and connectivity. Embeddedness refers to the mix of activities in each region through buyer supplier relationships. A region with more diversity mix can resist and react better to external shock and/or business cycle than a region without a diversified sector portfolio. Skilled-relatedness is related to the answers of the labour market to cycle fluctuations. It addresses the laid-off employee in relation to labor market absoption, i.e. if a sector close to his/her original employment (inter-sectoral labour mobility) is available or if a worker can commute to another region (interregional labour mobility.) In this chapter, the descriptive data permits to focus essentially on the regional condition of optimal labour supply defined as a rate of social benefit on total employment. A region is likely to be more resilience to downturns if its workforce is larger than its social beneficiaries.

In the next section, we are going to focus on six innovative regions in Sweden. The reason for selecting six regions (on 21 Swedish regions) is to select the three main agglomerations of the country and three other less innovative regions (defined respectively as "leading innovative" and "innovative follower" regions in Table 5.3).

5.3 Innovative Swedish Regions—NUTS 3

In the chapter, the overall definition of innovative region is a mix between evolutionary theories of technological change and the dynamics of regional system (Iammarino 2004: 5). It assumes that innovative regions combine three main functional dimensions: (i) absorption of new knowledge, technology and innovation for the adaptation to local needs; (ii) diffusion of innovations throughout all constituent parts of the regional social fabric to strengthen the existing knowledge base, and (iii) generation of new knowledge, technology and innovation. The classification NUTS 3 for the regions is a statistical nomenclature of territorial units allowing European comparisons. The level three or, NUTS 3, is a standard level in Swedish national statistics. It defines knowledge, technology and innovation in terms of the distribution of intramural R&D in its 21 Swedish counties (Table 5.2).

Table 5.2 Distribution of intramural R&D expenditures amongst the 21 Swedish counties, in millions of SEK

County—NUTS 3 level	All sectors	Share, %	Business enterprise sector	Higher education sector	Government sector
Stockholm County	45,026	36.1	33,030	10,265	1731
Västra Götaland County	25,144	20.2	18,814	5740	590
Skåne County	17,562	14.1	11,719	5292	551
Östergötland County	8884	7.1	6557	1720	607
Uppsala County	7413	5.9	2111	5160	140
Västerbotten County	3300	2.7	554	2509	237
Kronoberg County	2076	1.7	1808	248	20
Västmanland County	1866	1.6	1700	132	34
Örebro County	1660	1.3	1263	323	74
Dalarna County	1579	1.3	1405	111	64
Norrbotten County	1436	1.1	504	856	76
Jönköping County	1409	1.1	1189	193	27
Södermanland County	1305	1.0	1222	68	14
Gävleborg County	1207	1.2	1048	119	40
Västernorrland County	826	0.8	595	200	31
Värmland County	761	0.6	426	311	24
Halland County	646	0.5	497	111	38
Blekinge County	526	0.4	353	142	31
Kalmar County	476	0.4	295	168	13
Jämtland County	181	0.0	21	142	18
Gotland County	28	0.0	6	21	1
Not regionally distributed	1326	0.8	818		228

Source: SCB, Statistics Sweden, 2013

Table 5.3 Distribution of intramural R&D expenditures amongst the 21 Swedish counties, in 4 categories, 1- Driving Innovation, 2- Following Innovation, 3- Little Innovation, 4- No Innovation

Distribution of intramural R&D	14–37%	2–8%	1–2%	Less than 1%
21 Regions	Stockholm, Västra Götaland, Skåne	Östergötland, Uppsala, Västerbotten	Kronoberg, Västermanland, Örebro, Dalarna, Norrbotten, Jönköping, Södermanland, Gävleborg	Västernorrland, Värmland, Halland, Blekinge, Kalmar, Jämtland, Gotland

Source: Authors. One distinguishes leading innovative regions (14–37%) and following innovative regions (2–8%)

Notice, the percentage budget distribution of R&D funding over the regions is extremely skewed in favour of Stockholm. We decide, for the sake of comparison further to classify the regions into four categories to select the six most innovative ones (Table 5.3):

This chapter focuses on the two first categories of innovative regions (driving and following innovative regions). The selection of the six Swedish regions represents the most active industrial regions of the country. It is a well-known fact in regional studies that Sweden' innovative regions are concentrated around its three main metropolitan areas: Stockholm, Gothenburg, and Malmö. The capital region of Stockholm hosts many key learning institutions and concentrates most of capital and a large part of the industrial activities of the country. It concentrates 36.1% of the intra-mural R&D of the country. Gothenburg, located in Västra Götaland, hosts also important learning institution and a substantial share of industrial activities. It concentrates 20.2% of the intramural R&D. The Skåne region, whose main city is Malmö, the third largest city in Sweden, is hosting one of the most important universities of the country, Lund, with international research facilities in physics and bio-technology. Its intra-mural R&D activities represent 14.1%. Despite the average size of Malmö, the region is economically integrated within the Oresund region in the Baltic underlying intensified interactions with a major economic hub of Copenhagen, the capital of Denmark.

The other three regions we have selected along this list are Östergötland county, comprising the city of Linköping, which host an important university, and belongs to the most industrially active regions in the geographical crescent defined by Skåne country in the south, going through Västra Götaland and ending in Stockholm county. This country represents 7.1% share of the intramural R&D. Uppsala country is also known for its university of the same name and its closeness to Stockholm industrial areas. Its share of 5.9% of the intra-mural R&D makes it a regional player in innovation. Västerbotten county, the last in our list, is a very different player in this list. It has a substantively less importance in intramural R&D share than the other top regions in Sweden (2.7% which is half of Uppsala output). It has an important university in the city of Umeå. The region is characterized by its remoteness in the northern part of the country to other major economic hubs. The contrast between the two types of regions (driving and following innovative regions) will presumably help us to identify if there are clearly different pattern of employment resiliences between them.

As Martin (2012: 12) notes, the long-term adaptive capacity of regions is still 'largely un-researched'. The chapter looks at the decade 2004–2014 to learn how two types of innovative regions recess and grow from the job market point of view.

5.4 Some Exploratory Empirics: Labour Market's Accessibility, Dynamics and Performance

Let us start with an overview of the six regions in regard of total regional employment during the 10 years period (Figs. 5.1 and 5.2). First, let us notice that employment trend is continuously going upward in all six regions considered during the period.

Over the decade, both Stockholm and Uppsala regions have an employment growth of 21 and 18% respectively. We may consider, hypothetically, the existence of a correlation between those two regions (those are adjacent to each other's—Figs. 5.1 (Stockholm) and 5.2 (Uppsala) may explain their positive resilience to change. The other regions employment has grown less substantially, 13% for Skåne and 9% for the last 3: Västra Götaland, Östergötland and Västerbotten.

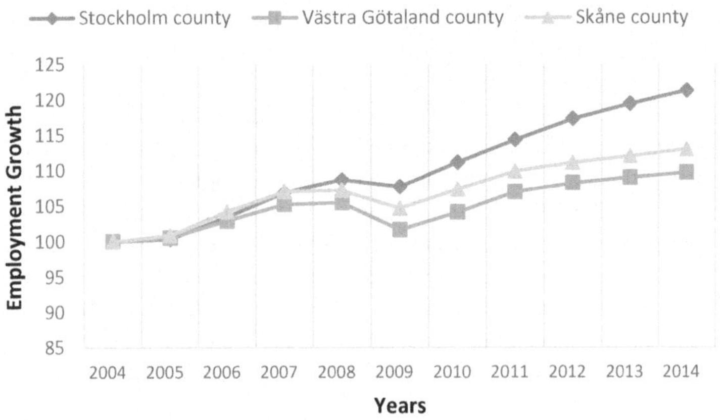

Fig. 5.1 Employment growth of the three leading innovation regions, yearly 2004–2014, indexed to 2004 = 100. Source: SCB, Statistics Sweden, 2015

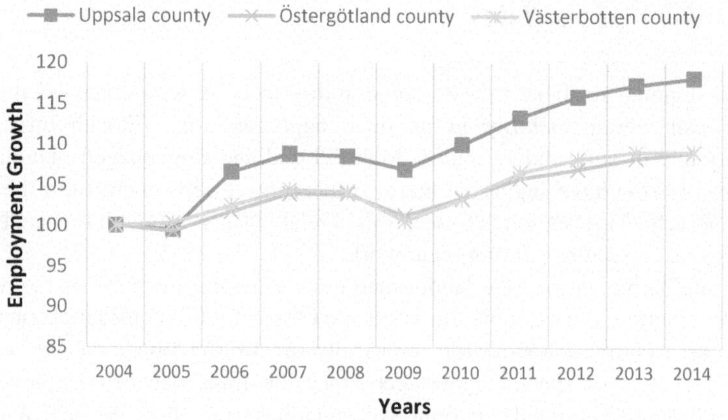

Fig. 5.2 Employment growth of the three following innovative regions, yearly 2004–2014, indexed to 2004 = 100. Source: SCB, Statistics Sweden, 2015

5.4.1 Employment Cycle or Shock?

All regions have experienced an inflection in employment rate in 2009. A shock is defined (Martin 2012: 15) by a longer time to recover (than production output) inflicted by a major decline in regional employment which consequences are profound for the region. For example, a shock in employment would take longer than industrial production to grow again. All statistical indication in the last 10 years in regional employment growth in Sweden stresses an inflection in economic cycle i.e. a modulation of investment in labour force which is followed by an immediate recovery (Ormerod 2010). In other words, we should talk of labour market sensitivity to economic cycle. Diodato and Weterings (2014: 20) talked of adaptive resilience when "a region offers laid-off workers to find a new job quickly, even if this means that they have to commute to other regions." Sweden labour adjustment is dependent on the global open economy notably with extensive foreign trade and its integration of international financial market.[2] It means that employment in Sweden follows the expansion and contraction of Swedish exports goods on the international market.

[2]The contraction in finance and credit on global market are acted by economic and government actors. Those adapting movements are common since Sweden's dependence on the international market fluctuation has increased over time. For example, Swedish export represents 44.56% of GDP in 2014. Market funding, not in the form of deposits, accounts for around 60% of banks' total balance sheets. Finance is acquired on international markets.

5.4.2 *Work Accessibility*

Swedish regional resilience shows some ability to cope with changes[3] thanks to interregional labour mobility in its main agglomerations (Stockholm-Uppsala; Göteborg and its surrounding region, Malmö/Lund and Copenhagen). Diodato and Weterings (2014) have suggested that one measure of adjustment from the labour force in a sector in recession is to seek job in related skills sets. Job seekers become mobile to reach sectors offering them work.

One alternative view to the labour market as a "shock absorber" is the regional networks function. For example, Karlsson and Olsson (2000) considered commuting patterns as a common source for empirically identifying functional regions. The labour market is of special importance since the links between employers and employees create a regional economic system (Johansson 1992). As such, a regional economic system is formed by interactive elements, and commuting is one of them. Those ways of working show economic relationship between regions but also organizational coupling between industrial life (working place) and individual choices (living place). Commuting patterns can display regional resilience as they organize continuity between actors and flows of goods and services. Let us observe how the six selected Swedish innovative regions define their job market accessibility through commuting.

Regionally, we find a commuting surplus if the incoming commuting flow is greater than the outgoing. If the outgoing and incoming commuting flows are equal, we reach a commuting equilibrium. Conversely, one finds a commuting deficit, if the outgoing commuting is greater than incoming. Two regions distinguish themselves as commuting surplus regions. Stockholm region and Västra Götaland (Gothenburg) have a large labour market (Fig. 5.3).

Stockholm region is showing an intensifying commuting surplus over the decade, with no sign of weakness. The raise in 2007, suggests that Stockholm offers commuters from other regions the labour adjustment they need in situations of economic slowdown. The following year 2008, shows no slowdown with a peak of commuter's surplus at 15%. Västra Götaland region is showing a variation of 4% in its pattern of surplus commuting. It represents around 400 people. The variation shows some regional adjustments of the labour market to changes in production and services.

The other four regions of our panel of innovative regions are all showing a decade of commuting deficit (Fig. 5.4). Not all regions show the same resilience to market change. The Fig. 5.4 is showing regional deficit by indicating negative values below the 100 indices.

One of the ways to explain regional resilience, i.e. the ability of a region to cope with change in labour market structure is to give an appreciation of regions' share of

[3]In contrast with the strict ability of regions to overcome shock by resuming pre-recession growth. The difference of interpretation is relative to the consideration of the length of historical cycle of regional growth.

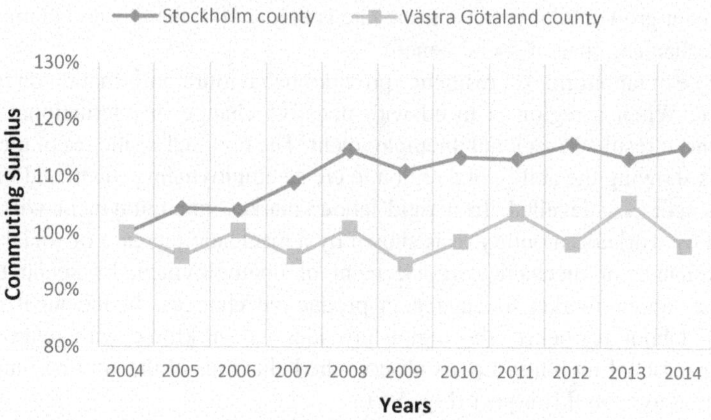

Fig. 5.3 Commuting surplus of 2 leading innovative regions, yearly 2004–2014, indexed to 2004 = 100. Source: SCB, Statistics Sweden, 2015

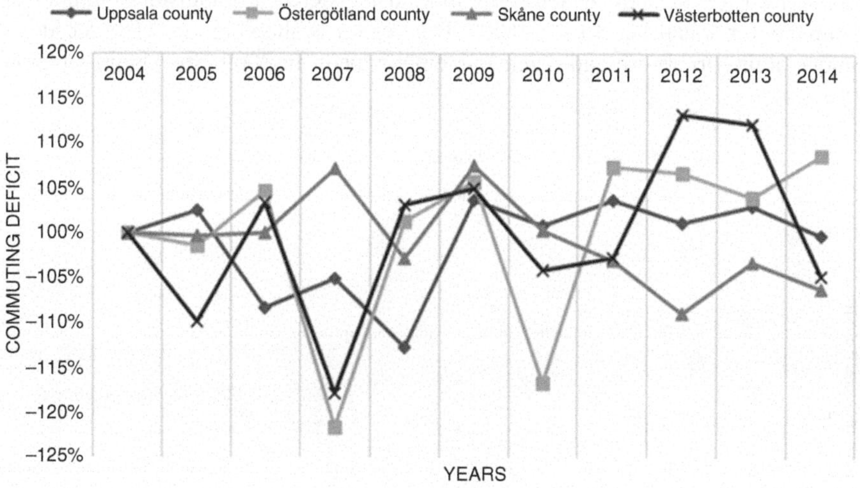

Fig. 5.4 Commuting deficit of 4 Swedish innovative regions, yearly 2004–2014, indexed to 2004 = 100. Source: SCB, Statistics Sweden, 2015. For three following innovative regions (Uppsala, Östergötland, Västerbotten) and one leading innovative region (Skåne)

gainful employment on social benefit (or "ungainful employment"). Social benefit is defined largely by including students, retired persons (whom total income is their pension), sickness beneficiaries and unemployment support. A labour market is resilient if it can afford to maintain optimal labour market conditions (low sickness, low health care cost, and low unemployment) in relation to social benefit. Regions will show labour resilience if their share of social benefit in comparison to employment is close to one (100% in the Fig. 5.5). If the rate is higher than 100, we have employment growth. Conversely, if the rate is below 100, there is no labour growth and a mechanical raise of social benefit.[4]

The role of labour market resilience provides behavioural information on regional resilience. When a region is faced with negative change in international market conditions, it results in regional unemployment. The regional resilience of the labour market is showing the ability of a region to create employment growth and maintain social benefit cost levelled. In a rigid labour market, unemployment will not be replaced by workers' mobility. It is shown by a mechanic increase of social benefit costs. Reliance of mechanical replacement of unemployment by social benefits limits the labour market to engage in productive changes. In the ideal case of "perfect" labour resilience, the region provides its workforce with related skills and inter-regional mobility. Let us observe the behaviour of the six first innovative Swedish regions on this aspect (Fig. 5.5):

Let us start with Stockholm region. Its results show a paradigmatic case of regional labour resilience. During the decade 2004–2014, its share of gainful employment follows a stable development. During the crisis of 2009, characterized by unemployment, Stockholm is the only innovative region registering an increase in employment (+ 31,000). In the same time, it registered a diminishing number of claimants for social benefit (– 4600). This region is showing, over the decade, a unique ability to absorb and regenerate employment. In 2009, Stockholm's region

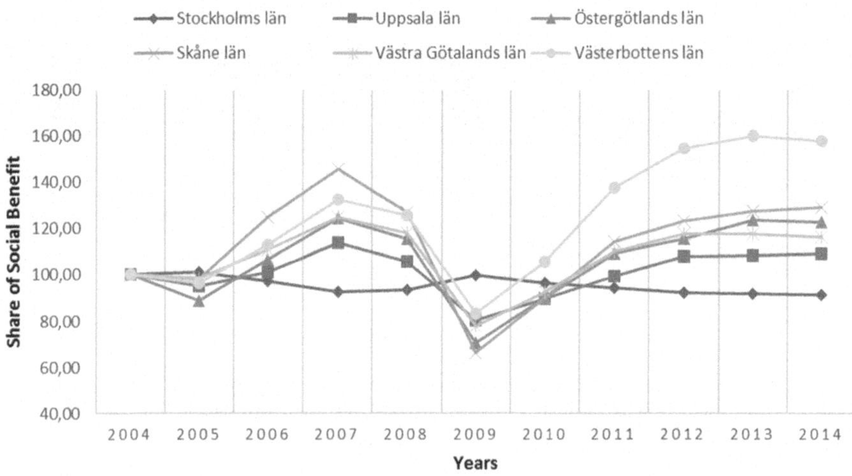

Fig. 5.5 Regional share of social benefit in the working population of 6 Swedish innovative regions, yearly 2004–2014, indexed 2004: 100. Source: SCB, Statistics Sweden, 2015

[4]We say "mechanical" rise of social benefit to reflect Swedish labour law (act 1997: 238). In Sweden, the welfare rules stipulate that people becoming unemployed have a right to claim 60 weeks of social benefit (1 year and 3 months) based on average salaries of your previous employment(s).

absorbed the surplus of available work in adjacent regions (and possibly the country).

The five other innovative regions—Västergötaland (Gothenburg), Skåne (Malmö-Lund), Uppsala region, Östergötland region (Linköping) and Västerbotten (Umeå) behaved similarly to each other during that decade. All have demonstrated a straightforward labour growth in 2007. In the case of Västergötland and Västerbotten, both regions had positive balance created by a straight labour growth accompanied with a diminishing social cost. In 2007, only Skåne had a strong labour growth with the same level of social benefit as the previous year.

All those five regions experienced a downturn of employment in 2009. In the Fig. 5.5, we can see the ratio of gainful employ turning negative (see the social benefit indicated below 100). All those regions—except Stockholm region—have registered a substantial employment drop. In the same time, they have mechanically distributed proportional social benefits. Let us notice that, in 2009, Skåne region is a paradigmatic case of labour rigidity. It faces unemployment (− 13,000) which brings mechanically a increased number of social benefit claims (+ 28,000). The 2009 labour shock is related to market function reflected in the temporary deficit of domestic or international demand. Those figures are not related to industrial regeneration. Therefore, competitive supply allows the recovery to take place for the following years to come.

5.4.3 Labour Market Efficiency

The section defines a complementary dimension of regional resilience by considering the region's ability to accommodate change toward continuous growth. In regional economics, a region is resilient when it absorbs employment crisis and permit further growth, i.e. when it maintains an efficient labour market. An efficient labour market is defined by an active control of public expenditure to allocate opportunity costs to market orientated process (Demmke and Moilanen 2012). In addition, efficient labour market creates the conditions of regional resilience when resources are allocated to respond effectively to the changing needs in society, i.e. the ability to reallocate skills in adjacent sectors or regions (Behrenz et al. 2013). Accordingly, a regional labour market is efficient if the share of public employment is lower than private employment to drive growth. Conversely, a labour market is stagnant if the share of public employment is growing continuously in proportion of private employment. The role of labour market dynamism in regional resilience is showing the ability of region to absorb employment crisis by generating conditions for labour market rebound. The regional dynamism of the labour market is showing the ability of a region to engage in policies of labour adjustment through access opportunities (either in related industrial sectors or through inter-regional mobility). Let us consider some descriptive empirics on the behaviour of the six first innovative Swedish regions (Fig. 5.6):

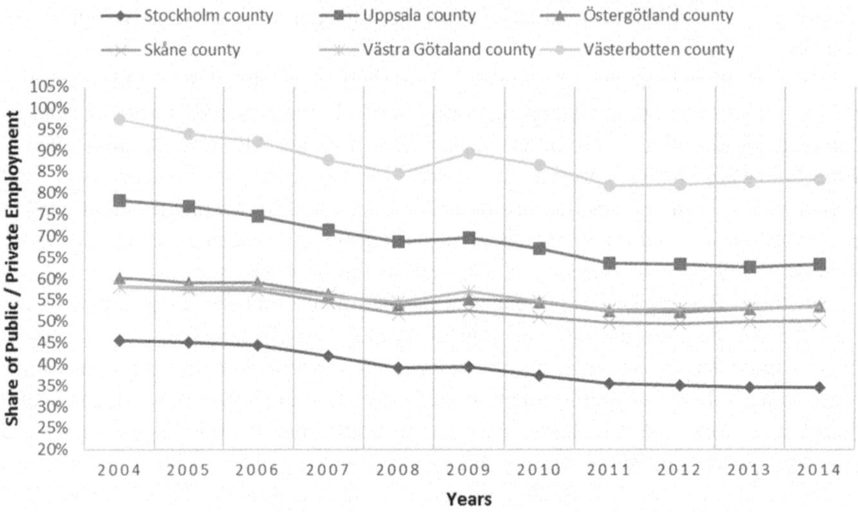

Fig. 5.6 Public/private employment growth rate in the six selected Swedish regions 2004–2014. Those data excludes employment from other non-private or public organizations. Source: SCB, Statistics Sweden, 2015

In 2004, the share of public sector in the private sector employment is of 46% in Stockholm region. The most innovative region of Sweden is also the region in which the control of public employment is the most rigorous. During the decade, the region reaches a peak of efficiency with the lower rate of 34% in 2014. Since 2004, the administrations[5] employ roughly the same number of employees from 272,155 in 2004 to 272,554 in 2014. The employment dynamics reflects the surplus of the private sector employment. This sector created 598,047 employments in 2004 to 793,070 ten years later.

The other five innovative regions do not follow such a positive pattern. Three of them, Västra Götaland (Göteborg), Östergotland (Linköping) and Skåne (Malmö/Lund) have similar employment behaviour. Their public-sector employment represents 60% of the total employment in 2004. During the decade, all of them have work to control this ratio moving toward a balanced 50/50 (53% for Västra Götaland and Östergotland). The detail indicates that Skåne worked in controlling its public-sector employment. For example, the region had, from the year 2008 to 2011, the same number of public employees than 2004. The region has registered a general increase of 10,000 public sectors employees over the decade. The private sector has

[5]In our data, the administration includes central government, central government quasi-corporations, primary local government, county councils, other public institutions, central government corporations and organizations, local government corporations and organizations. The private institutions are joint-stock corporations not controlled by the government sector. Other corporations not controlled by the government sector. Our data do not consider "other organizations". We have noticed the number of "other organizations" in all regions is significantly stable.

succeeded in creating a positive dynamic in raising employment by 62,000 during the decade.

Västra Götaland and Östergötland have also a similar public/private number of employees' rate around 60% in 2004. Both regions are controlling this rate by diminishing it by 1% a year reaching a balanced average of 50% (and 53 respectively). Both counties have grown their public sector by 2% and 4% respectively. Their private sector has grown by 15% and 14% respectively.

In 2004, Uppsala and Västerbotten counties show a highest rate of public employment in innovative regions, scoring around 78% and 97% respectively. Both have shown willingness to control an over-administered social fabric by reducing it slowly during a decade to 63 and 83% respectively. Both public and private sectors in Uppsala county have grown by 5% showing no employment dynamism. The Västerbotten region is showing the political prominence of subsidies related to its remote geographical positioning (localized in the north of Sweden and remote from the three major agglomeration economies of the country—Stockholm, Gothenburg, and Malmö). Over the decade 2004–2014, its private sector dynamism is growing but at a relatively low rate (16%). This region is highly dependent on public funding. Although the region is dependent on it, it controls its spending allowing a 2% growth over the whole decade. Västerbotten is a paradigmatic case of a lack of labour market dynamism. The challenge of this region is to create profit since the number of employees in the public and private sector is almost identical.

5.5 Conclusion

This chapter is conceptualizing the notion of regional resilience without emphasizing recessionary shocks and other long-term disruption. "Shock theories" are based on data set from the 1970s and 1980s onward reporting industrial downturn of that period. Our descriptive empirics are based on the last 10 years which reports common fluctuations in business cycles. From 2004 to 2014, regions display heterogeneous dynamics in regard of their employment market. Those differences affect our way of theorizing regional resilience. In this chapter, "resilience" is adaptive as far as we consider region's ability to adapt to socio-economic trends. This chapter confirms the basic knowledge in Swedish regional economics, that innovation is skewed centrally toward the Stockholm region. The chapter's selection of six innovative regions adopts a voluntarily loose definition (leader, follower) of economic indicators for innovation. This choice allows us to question the idea of regional resilience by shifting our attention away from measures for improving innovative competitiveness. Our suggestion is to consider, in line with policy consideration, resilience from the point of view of the structure of its human resources, i.e. labour dynamics and employment. Our selections of labour descriptive empirics focus on region resilience defined by three main indicators: (1) its accessibility as commuting in and out of the region; (2) employment resilience defined as the share of employment to social benefit and (3) employment market efficiency defined as the share of public employment in total employment. As eluded before, the study shows that Stockholm regions are a major employment hub, with positive employment resilience and an effective employment market. Västra Götaland attracts regional commuters and has employment resilience sensitive to employment downturns. However, the region is dynamic showing recovery and growth thanks to positive labour market efficiency. Skåne endorses some problematic characteristics. Despite a negative commuting pattern, the regions benefit from providing some of its work force to Denmark's capital, Copenhagen. What matter the most is Skåne's rigid employment market. It shows an abnormally large use of social benefit in time of employment downturn. This greater reliance on social benefit reflects limitations in the ability to access the labour market through sector or regional mobility. Despite this aspect, the region has a high potential for labour development. The other three regions—Östergöteland, Uppsala and Västerbotten are interesting contrasting cases for three reasons:

1. Their development over the last decade shows a similar resilience pattern defined by labour growth and an ability to absorb employment downturn.
2. Those regions stand out by the limitation of their labour market's adaptability. Their employment depends upon larger metropolitan centres. They possess the ability to take advantage of their complementarity with more powerful adjacent regions, but are less resilient compared to the leading innovative regions when employment downturn hits them. Further, they do not have the capacity to generate their own capacity for employment.
3. Those regions have a labour market more sensitive to their regional idiosyncrasies (Bristow 2010). This questions the extent to which those regions can pilot their employment downturn, exploit their labour mobility and develop sustainable growth from their level alone.

5.6 Discussion

The chapter proposed an evolutionary view of regional resilience based on three levels of labour behaviour (accessibility, social benefit and labour market efficiency) which opened a whole set of new research challenges. In the following, we briefly discuss a few of them. Our concept of regional resilience is geared toward policy implication by focusing on labour dynamics. The first conclusion we draw is that all Swedish regions are resilient in terms of employment over a decade. The over-dramatizing story of "disrupting shock from the market" does not hold since innovative regions are able to absorb changes in the market over a year period. Innovative regions seem to benefit from dynamic economic conditions. The contrast between less endowed regions in terms of innovation shows that less innovative regions are struggling to create conditions for growth. It is clearly due to a lack of employment mobility, an important reliance of social subsidies and, too little sustainable entrepreneurship (a large share of public employment over private.). Let us specify below few points needing further discussion.

First, this preliminary study confirms that innovative regions (Stockholm & Västra Götaland) are resilient based upon the labour indicators of high labour accessibility, high employment recovery combined with low social benefit and a significant surplus of employment in the private sector. One expects innovative regions with diverse industrial make-up (such as a variety of skill-related industries) to overrun "bumps" of economic downturn but also to create suitable growth conditions.

Second, the contrast with less innovative regions questions the condition for resilience. Not all the other regions outside Stockholm and Västra Götaland have a positive commuting pattern. It suggests that those regions are less equipped to deal easily with economic change (downturn or growth) given reduced sectors diversity. Here, less innovative regions (in our Table 5.2-2 the category 2 of "innovation followers") have less human resources to growth. Although we notice that all those regions have worked on diminishing their social benefit over a decade, research policy may investigate complementary policy to support employment growth. For example, more radical taxation exemption scheme to create condition for creating or attracting new firms is one way to boost employment.

The third and last contribution is related to the abilities of those regions to grow. The contrast between the innovative and less innovative regions in the sample is telling. The question of regional growth asks the more fundamental question of the ability of regions to drive growth, i.e. to create a positive surplus of successful businesses and industries. Stockholm region is the only one who scores high on employment growth, showing all positive indicators of regional resilience. It has a low share of public sector employment in its total employment (34% now) as a resulting tendency to reduce that rate in the last 10 years. Other innovative regions, such as Västra Götaland, Östergötland and Skåne have a high level of public employment moving toward a healthier balance ratio 50/50 between public and private sectors (60% in 2004 going toward 50% 10 years later). Further research should seek to know if this balance toward the private employment is a necessary ground of a dynamic labour diversity (European Commission, COM 2014). The study shows that contrast between regions relative to their level of innovation is essential to their ability to absorb downturn successfully and positively (economic rebound). Most regions in Sweden, and the few categorized as "innovating followers" such as Uppsala and Västerbotten have a very high public employment rate. They are also showing similar reduction of public employment in their total employment in the last 10 years. However "lower innovative regions" have clearly different initial conditions than innovative ones.

This study brings several policy implications. In this perspective, the results suggest a need to shift focus from the resilience-to-industrial chock to the flexibility of human resources in economic cycles. The resilience of Swedish regions challenges our ability to conceive better labour accessibility and diversity in regions with lower innovative endowment. In regional economics, it is often argued that the agglomeration allows creative solutions such as knowledge spillovers, industrial combination and institutional overlap (Boschma 2005). From a labour perspective, regional accessibility and diversity is likely to emerge from young industries in economically attractive regions (Capello et al. 2011). The review of a decade of regions' growth in Sweden questions the theory of the "recurrent crisis of capitalism", through recovery and reconversion from industrial shock (Martin 2012) as well as recession from the 2009 crisis (Martin et al. 2015). Concerning Sweden, the alarming stance of the "endemic crisis of capitalism" is a theme of the past. Regional resilience is a meso-level phenomenon demanding a specific attention at the junction between regional labour markets and human capital management. Regional policies should work on regional attractiveness according to the following points:

1. Innovative regions are competing nationally and internationally on continuous flows of economic factors and financial capital. Less innovative regions clearly do not. In Sweden, few regions can afford to be worldwide players.
2. The resilience of labour shows that regions controlling their social expenditure can increase marginally their productivity, or employment growth. Both combinatory policies and political risk will decide if it is possible to bring less innovative regions into intra-regional and worldwide competitiveness. This chapter shows that in Sweden, less innovative regions rely more on subsidized social benefits and a larger public sector.
3. In western countries, where infrastructure exists, the issue of creating diversified and specialized new activities demand policy maker to challenge their planning bend in favour of the market tested betterment. New industries are hardly created by policy driven initiatives.
4. The abilities of regions to generate new businesses or alternatively to simplify the establishment of new businesses would create the condition for regional regeneration. Flexible tax law, labour policies and international trade facilitate unrelated diversification, i.e. the ability of the work force to move seamlessly into new fields creating regional condition for growth.

References

Almeida, P., & Kogut, B. (1999). Localization of knowledge and the mobility of engineers in regional networks. *Management Science, 45*(7), 905–917.

Amable, B. (2000). Institutional complementarity and diversity of social systems of innovation and production. *Review of International Political Economy, 7*(4), 645–687.

Andersson, M., & Karlsson, C. (2004). *Regional innovation systems in small and medium sized regions. A critical review and assessment* (Working paper series in economics and institutions of innovation, no. 10). Stockholm: CESIS Royal Institute of Technology Stockholm.

Andersson, M., & Koster, S. (2011). Sources of persistence in regional start-ups. Evidence from Sweden. *Journal of Economic Geography, 11*(1), 179–201.

Asheim, B., & Isaksen, A. (2002). Regional innovation systems: The integration of local sticky and global ubiquitous knowledge. *The Journal of Technology Transfer, 27*(1), 77–86.

Balland, B. A. (2012a). Proximity and the evolution of collaboration networks: Evidence from research and development projects within the global navigation satellite system (GNSS) industry. *Regional Studies, 46*(6), 741–756.

Balland, B. A. (2012b). *Promoting knowledge transfer and the economic resilience of regions* (Working paper). Utrecht: Utrecht University.

Behrenz, L., Delander, L., & Månsson, J. (2013). *The public employment service: A survey of cost-benefit analysis and productive efficiency analysis studies*. Växjö: Linnéuniversitetet.
Boschma, R. (2005). Proximity and innovation: A critical assessment. *Regional Studies, 39*(1), 61–74.
Boschma, R. (2015). Towards an evolutionary perspective on regional resilience. *Regional Studies, 49*(5), 733–751.
Boschma, R., & Frenken, K. (2010). The spatial evolution of innovation networks. A proximity perspective. In R. Boschma & R. Martin (Eds.), *The handbook of evolution economic geography* (pp. 120–135). Cheltenham: Edward Elgar.
Boschma, R., & Lambooy, J. (1999). The prospects of an adjustment policy based on collective learning in old industrial regions. *Geojournal, 49*, 391–399.
Boschma, R., Minondo, A., & Navarro, M. (2013). The emergence of new industries at the regional level in Spain. A proximity approach based on product-relatedness. *Economic Geography, 89* (1), 29–51.
Bristow, G. (2010). Resilient regions: Re-'place'ing regional competitiveness. *Cambridge Journal of Regions, Economy and Society, 3*, 153–167.
Caballero, R., & Hammour, M. (1994). The cleansing effect of recession. *American Economic Review, 84*, 1350–1368.
Capello, R., Fratesi, U., & Resmini, L. (2011). Regional attractiveness and its determinants. In *Globalization and regional growth in Europe* (pp. 191–213). Berlin: Springer Verlag.
Cellini, R., & Torrisi, G. (2014). Regional resilience in Italy: A very long run analysis. *Regional Studies, 48*, 1779–1796.
Christopherson, S., Michie, J., & Tyler, P. (2010). Regional resilience: Theoretical and empirical perspectives. *Cambridge Journal of Regions, Economy and Society, 3*, 3–10.
Clark, J., Huang, H. I., & Walsh, J. P. (2010). A typology of "innovation district": What it means for regional resilience. *Cambridge Journal of Regions, Economy and Society, 3*, 121–137.
Cooke, P., Bristow, G., & Porter, J, (2011, September). *Regional resilience literature review*. Paper presented Bordeaux
Davies, S. (2011). Regional resilience in the 2008–2010 downturn: Comparative evidence from European countries. *Cambridge Journal of Regions, Economy and Society, 4*, 369–382.
Demmke, C., & Moilanen, T. (2012). *The future of public employment in central public administration restructuring in times of government transformation and the impact on status development*. Maastricht: European Institute of Public Administration.
Diodato, D., & Weterings, A. B. (2014). The resilience of regional labour markets to economic shocks: Exploring the role of interactions among firms and workers. *Journal of Economic Geography, 15*(4), 723–742.
Doussard, M., & Schrock, G. (2015). Uneven decline: Linking historical patterns and processes of industrial restructuring to future growth trajectories. *Cambridge Journal of Regions, Economy and Society, 8*(2), 149–165.
Ebbinghaus, B. (2009). Can path dependence explain institutional change? Two approaches applied to welfare state reform. In L. Magnusson & J. Ottosson (Eds.), *The evolution of path dependence* (pp. 191–218). Cheltenham: Edward Elgar.
Essletzbichler, J. (2015). Relatedness, industrial branching and technological cohesion in US metropolitan areas. *Regional Studies, 49*(5), 752–766.
European Commission. (2014). Taking stock of the Europe 2020 strategy for smart, sustainable and inclusive growth, COM (2014), 130 final, Brussels.
Fingleton, B., Garretsen, H., & Martin, R. (2012). Recessionary shocks and regional employment: Evidence on the resilience of UK regions. *Journal of Regional Science, 52*, 109–133.
Fingleton, B., Garretsen, H., & Martin, R. (2015). Shocking aspects of monetary union: The vulnerability of regions in Euroland. *Journal of Economic Geography, 15*(5), 907–934.
Foster, K. A. (2007). *A case study approach to understanding regional resilience* (Working paper 2007–2008). Berkeley, CA: Institute of Urban and Regional Development.

Graddy-Reed, A., & Feldman, M. P. (2015). Stepping up: An empirical analysis of the role of social innovation in response to an economic recession. *Cambridge Journal of Regions, Economy and Society, 8*, 293–312.

Grillitsch, M. (2014). *Institutional change and economic evolution in regions* (Working paper). Lund: Lund University.

Hall, P. A., & Soskice, D. (Eds.). (2001). *Varieties of capitalism. The institutional foundations of comparative advantage*. New York: Oxford University Press.

Hassink, R. (2010). Regional resilience: A promising concept to explain differences in regional economic adaptability? *Cambridge Journal of Regions, Economy and Society, 3*, 45–58.

Henning, M., Stam, E., & Wenting, R. (2013). Path dependence research in regional economic development: Cacophony or knowledge accumulation? *Regional Studies, 47*(8), 1348–1362.

Hill, E., Wial, H., & Wolman, H. (2008). *Exploring regional economic resilience* (Working paper 2008–04). Berkeley, CA: Institute of Urban and Regional Development.

Hill, E., St Clair, T., Wial, H., Wolman, H., Atkins, P., Blumenthal, P., et al. (2012). Chapter 6: Economic shocks and regional economic resilience. In N. Weir, N. Pindus, H. Wial, & H. Wolman (Eds.), *Building resilient regions: Urban and regional policy and its effects* (Vol. 4, pp. 193–274). Washington, DC: Brookings Institution Press.

Hollingsworth, J. R. (2000). Doing institutional analysis: Implications for the study of innovations. *Review of International Political Economy, 7*(4), 595–644.

Iammarino, S. (2004, June 4–5). *On the definition of regional system of innovation (RSI): An application to the Italian case*. Conference on "Regionalisation of Innovation Policy", Berlin, Germany.

Johansson, B. (1992). *Ekonomisk Dynamik i Europa*. Malmö: Liber-Hermods.

Johansson, B. (1993). Economic networks and self-organization. In E. Bergman, G. Maier, & F. Tödtling (Eds.), *Regions reconsidered*. Suffolk: Ipswich Book Co Ltd.

Johansson, B. (1998). *Infrastructure, market potential and endogenous growth*. Jönköping: JIBS.

Magnusson, L., & Ottosson, J. (Eds.). (2009). *The evolution of path dependence*. Cheltenham: Edward Elgar.

Markusen, A. (1985). *Profit cycles, oligopoly, and regional development*. Cambridge, MA: MIT Press.

Martin, R. (2010). Roepke lecture in economic geography. Rethinking regional path dependence. Beyond lock-in to evolution. *Economic Geography, 86*, 1–27.

Martin, R. (2012). Regional economic resilience, hysteresis and recessionary shocks. *Journal of Economy Geography, 12*, 1–32.

Martin, R. L., & Sunley, P. J. (2015). On the notion of regional economic resilience: Conceptualisation and explanation. *Journal of Economic Geography, 15*, 1–42.

Martin, R., Sunley, P., & Tyler, P. (2015). Local growth evolutions: Recession, resilience and recovery. *Cambridge Journal of Regions, Economy and Society, 8*(2), 141–148.

McCann, P., & Ortega-Argilés, R. (2013). Modern regional innovation policy. *Cambridge Journal of Regions, Economy and Society, 6*(2), 187–216.

Meeus, M., Oerlemans, L., & van Dijck, J. (1999). *Regional systems of innovation from within – an empirical specification of the relation between technological dynamics and interaction between multiple actors in a Dutch region*, ECIS working paper no. 99.1.

Neffke, F., Henning, M., Boschma, R. A., Lundquist, K. J., & Olander, L. O. (2011a). The dynamics of agglomeration externalities: A survival analysis. *Journal of Economic Geography, 12*(2), 485–517.

Neffke, F., Henning, M., & Boschma, R. (2011b). How do regions diversify over time? Industry relatedness and the development of new growth paths in regions. *Economic Geography, 87*, 237–265.

OECD. (2012a). What makes labor markets resilient during recessions? In *OECD employment outlook 2012*. Paris: OECD Publishing.

OECD. (2012b). What makes labor markets resilient during recessions? In *Annexes 2.A1 and 2.A2 of chapter 2 of the 2012 OECD employment outlook*. Paris: OECD Publishing.

OECD. (2012c). *Going for growth: Economic policy reforms*. Paris: OECD Publishing.
Ormerod, P. (2008). Resilience after local economic shocks. *Applied Economic Letters, 17*, 503–507.
Ormerod, P. (2010). Risk, recessions and the resilience of the capitalist economies. *Risk Management, 12*(1), 83–99.
Pike, A., Dawley, S., & Tomaney, J. (2010). Resilience, adaptation and adaptability. *Cambridge Journal of Regions, Economy and Society, 3*, 59–70.
Reggiani, A., De Graaf, T., & Nijkamp, P. (2002). Resilience: An evolutionary approach to spatial economic systems. *Networks and Spatial Economics, 2*, 211–229.
Saviotti, P. P. (1996). *Technological evolution, variety and the economy*. Cheltenham: Edward Elgar.
Simmie, J., & Martin, R. (2010). The economic resilience of regions: Towards an evolutionary approach. *Cambridge Journal of Regions, Economy and Society, 3*, 27–43.
Strambach, S. (2010). Path dependency and path plasticity. The co-evolution of institutions and innovation – the German customized business software industry. In R. A. Boschma & R. Martin (Eds.), *Handbook of evolutionary economic geography* (pp. 406–431). Cheltenham: Edward Elgar.
Strambach, S., & Klement, B. (2012). Cumulative and combinatorial micro-dynamics of knowledge. The role of space and place in knowledge integration. *European Planning Studies, 20*(11), 1843–1866.
Swanstrom, T., Chapple, K., & Immergluck, D. (2009). Regional resilience in the face of foreclosures: Evidence from six metropolitan areas. In *A report prepared for the MacArthur foundation's building resilient regions project* (Working paper 2009–05). UC Berkeley: Institute of Urban and Regional Development.
Treado, C. D. (2010). Pittsburgh's evolving steel legacy and the steel technology cluster. *Cambridge Journal of Regions, Economy and Society, 3*, 105–120.
Wiig, H. (1996). An empirical study of the innovation system in Finnmark, STEP working paper R-09, Oslo.
Wolfe, D. A. (2010). The strategic management of core cities: Path dependence and economic adjustment in resilient regions. *Cambridge Journal of Regions, Economy and Society, 3*, 139–152.
Zeitschrift fur Wirtschaftsgeographie. (2013). Special issue on reconceptualizing change: Path dependency, path plasticity and knowledge combination 57(1–2), 1–96.
Zolli, A., & Healy, A. M. (2012). *Resilience. Why things bounce back*. London: Headline Publishing Group.

Chapter 6
Diversifying Mediterranean Tourism as a Strategy for Regional Resilience Enhancement

André Samora-Arvela, Eric Vaz, João Ferrão, Jorge Ferreira, and Thomas Panagopoulos

6.1 Introduction

Tourism-based regions, especially in southern Europe, are extremely vulnerable to the expected impacts of climate change (Amelung and Moreno 2009, pp. 16–25). In the case of the south of Portugal, the Algarve region's socio-economic base is grounded on summer sun and beach tourism, i.e. coastal tourism with large-scale and concentrated accommodation. This could be drastically affected by climate change, essentially derived from sea level rise and potential beach area reduction, constraining the future of a region that strongly depends on this type of tourism.

From the assumed weakness of the touristic use of this Mediterranean climate region to the impacts of climate change, it urges a reflection about what adaptive strategies can be considered.

There are several strategies that have already been implemented in other Mediterranean destinations, which should be taken into account. These constitute a transformative opportunity for change in the practice of tourism in the Algarve region, namely by the sustainable use of green infrastructures and a new look to

A. Samora-Arvela (✉) · J. Ferreira
Interdisciplinary Centre of Social Sciences, Faculty of Social Sciences and Humanities, New University of Lisbon, Lisbon, Portugal
e-mail: jr.ferreira@fcsh.unl.pt

E. Vaz
Department of Geography and Environmental Studies, Ryerson University, Toronto, Canada
e-mail: evaz@ryerson.ca

J. Ferrão
Institute of Social Sciences, University of Lisbon, Lisbon, Portugal
e-mail: joao.ferrao@ics.ul.pt

T. Panagopoulos
Research Centre for Spatial and Organizational Dynamics, University of Algarve, Faro, Portugal
e-mail: tpanago@ualg.pt

H. Pinto et al. (eds.), *Resilience and Regional Dynamics*, Advances in Spatial Science, https://doi.org/10.1007/978-3-319-95135-5_6

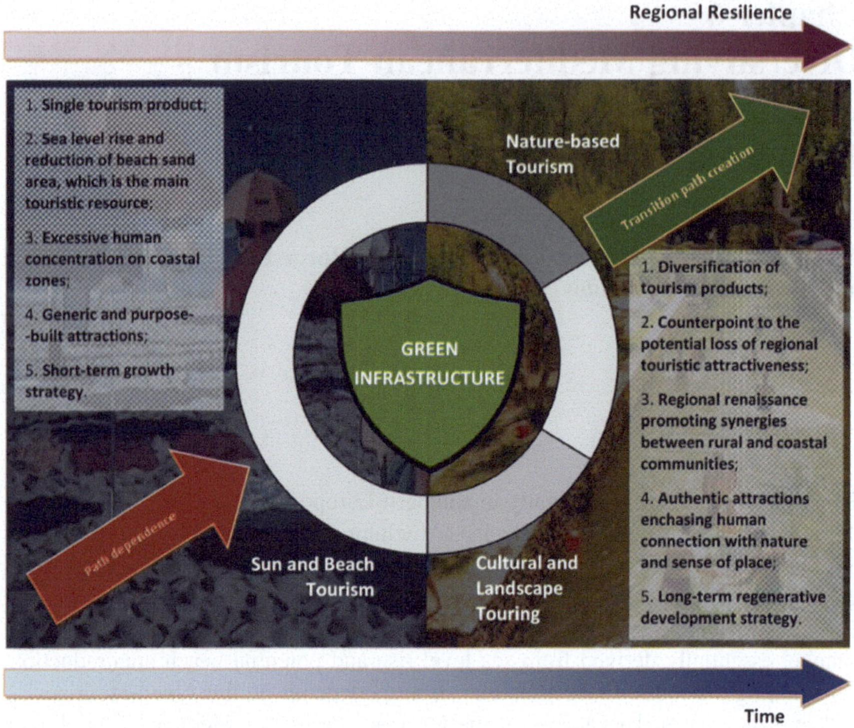

Fig. 6.1 Approach framework to the interrelationship between green infrastructure and climate change in Algarve region

its natural and cultural landscape units with small and dispersed accommodation, low seasonality and low impact recreation activities. In doing so, activities that could contribute to diversification and resilience enhancement of this touristic area, particularly in light of climate change, should be considered (Fig. 6.1).

The fundamental purpose of this chapter is to analyze how several Mediterranean destinations have transitioned to a green infrastructure approach. These approaches are examined for their sustainable tourism potential in order to stimulate the diversification and tourism competitiveness of the Algarve region. The integration of these approaches into current spatial planning policies and landscape management could be a means of adapting to present and future climate challenges.

6.2 Mediterranean Touristic Areas and Climate Change: Sustainability or Unsustainability?

Tourism practices existed in earlier times, such as the *Grand Tour* of the historical aristocracies, and travel for health therapy, amongst others. However, the tourism massification came only in the second half of the twentieth century due to the improvement of roads and transport, as well as the increase of leisure time and the greater financial availability of a large part of the population (Silva 2009, p. 25).

The same author also notes that tourism was, at first, seen as a fully candid and beneficial economic activity. However, from the 1970s, this vision was blurred by the perception of the negative effects of mass tourism development in the physical and social environment, cultural heritage and host communities, "as well as by finding that they did not receive the economic dividends gained from the tourist industry" (Silva 2009, p. 25). Here, tourism takes on a form of imperialism and of the capitalist commodification of culture, which has a destructive effect on it (Silva 2009, p. 25).

In this context, the term *Sustainable Tourism* emerged and was formulated about two decades ago. Research and publications on this subject began to intensify starting in 2008 (Zolfani et al. 2015).

Liu (2003) defines sustainable tourism as all forms of tourism that contribute to sustainable development and, therefore, trigger both sustainable economic and societal development. It is also defined by an efficient allocation of biophysical and socio-cultural resources, which can only be achieved on the basis of sound knowledge and careful management of tourist demand.

Weaver (2006, p. 10) presents the notion of sustainable tourism as derived from sustainable development concepts, namely that tourism development should meet the needs of the present without compromising the ability of future generations to meet their own. Thus, tourism development requires the conservation of resources (e.g. their long-term maintenance), while simultaneously minimizing the negative impacts and maximizing those that are positive.

Bramwell and Lane (2012) acknowledge that tourism, because of its significant growth, has become jeopardizing to sustainable development directives. A systematic reorientation towards the preservation and integration of the global environment must, therefore, be considered within the strategic policies for regional tourism.

In this sense, an alternative to conventional tourism has emerged called *Ecotourism.* According to Dernoi (1981), this alternative tourism differs from its mainstream counterpart in its greater pursuit of sustainable use of resources in regions, where tourists are greeted at home or on the property of local hosts (Weaver 2006, p. 39).

As a proxy of sustainable tourism, ecotourism was theorized in the 1980s, a time when Hector Ceballos-Lascuráin (1988, p. 13) defined it as travelling to natural-undisturbed or little-changed areas in order to experience the existing cultural heritage and study, appreciate or enjoy the aesthetic quality of the flora and fauna. It is therefore based on two criteria, namely nature-based and educational, or contemplative motivations (these are prerequisites of the ecotourism experience). Weaver (2006, p. 192) also quotes Epler Wood (Wood and House 1991, p. 200), who defined ecotourism as "the journey in order to know the cultural and natural history of an area, attempting to not alter the integrity of the ecosystem while producing, simultaneously, both economic opportunities for the conservation of natural resources, financially beneficial to local citizens". Fennell (1999, p. 43) considered ecotourism as a sustainable and nature-based activity that focuses on learning and on the connection to nature and landscape—an activity managed ethically to be both non-consumptive and locally-oriented, and to have a low impact, overall (Weaver 2006, p. 192).

Mass Tourism is characterized by a number of criteria, including: the tourists' brief length of stay; distinct seasonality between the high and low seasons; attractions that are generic and purpose built; large-scale and concentrated accommodation; architectural international style; non-local and corporate ownership, and linkages with non-local sectors (with an emphasis of profit and economic growth within a short-term timeframe). Comparatively, *Alternative Tourism* is defined by: an extended length of stay with no distinct seasonality; pre-existing and authentic attractions; small-scale and dispersed accommodation; a vernacular architectural style; the involvement of local sectors, and a focus on long-term strategy of community well-being (Weaver 2006, p. 41).

It is therefore important to reformulate tourism products so that the environment is preserved, the local culture and production is maintained, and the benefits of tourism with host communities are shared with key stakeholders.

It is true that "green", "sustainable" and "eco-friendly" are very common terms in sustainable tourism planning and research, but more needs to be done (Bowman 2011). Sharpley (2000) points out that the principles of sustainable tourism are difficult or almost impossible to achieve, but they are important to encourage more benign forms of tourism. Worthwhile, Zolfani et al. (2015) presents, two examples of such ecotourism: rural and cultural tourism. The incentives of these two examples should be planned in a climate change framework and in the context of combined mitigation and adaptation planning (Biesbroek et al. 2009; Scott 2011; Sharpley 2000).

The last and primordial need for this "smart development, spatial sustainability and environmental quality" in the "New Urban World" (Kourtit 2017) in tourism ground, emerges, fundamentally, through smart specialization that is supported by innovation, creativity, information and interaction (Romão and Neuts 2017). Smart investments in human and social capital, as well as in traditional and modern communication infrastructure, catalyze sustainable development, boosting high quality of life with an efficient management of natural resources, through participatory governance (Romão and Neuts 2017).

It is paramount to ensure the sustainable development of regions in their uniqueness and diversity of territorial resources and in the differentiation of tourism supply. This is critical so as to ensure both the delivery of high quality of tourism products and services as well as to ensure the preservation of the region (Romão and Neuts 2017) and its resilience to challenges in the future (Boschma 2015, p. 736; Christopherson et al. 2010).

In addition, the greatest challenge to be reflected in the context of evolving tourism is climate change, a strong obstacle to sustainable practices/regional development. This challenge reveals the pressing need to identify cause-impact relationships and to clarify the intricate roles of spatial planning and tourism in order to present solutions (Biesbroek et al. 2009; Scott 2011).

For this reason, it is not surprising that, in the present and future context, climate change represents a disruption element for this vulnerable sector and to the socio-economic basis of regions that depend on it (UNWTO 2008, pp. 61–68). The direct impacts expected include the redefinition of periods of climate pleasantness and respective seasonality of tourism demand, as well as changes in the operating costs of tourist establishments (Casimiro et al. 2010, pp. 6–7). The indirect environmental impacts expected include temperature increase, sea level rise and boosting occurrence of extreme events (e.g. heat waves, intense precipitation events, floods and droughts, which will cause a scarcity of water resources, coastal erosion and reduction of the sand beach area). Other potential indirect impacts include increasing the number of fires, decreasing agricultural production, a reduction of snow surfaces in winter sport destinations, landscape aesthetic quality degradation and an increased incidence of vector-borne diseases due to the disappearance of climate barriers, among others (Simpson et al. 2008, p. 13; Becken and Hay 2007, pp. 38–50; UNWTO 2008, pp. 61–68; IPCC 2014a, p. 1283).

The PESETA I European project foresees a deterioration of climate pleasantness in summer and its increase in the spring and autumn in Mediterranean climate destinations (Amelung and Moreno 2009, pp. 16–25). This foresight comes through the modelling of the Tourism Climate Index (TCI), which defines climate suitability for tourism under climatic factors, such as temperature, humidity, sunshine hours, precipitation and wind. Such predicted deteriorations may imply changes in the geography of tourist attractiveness, corresponding to societal impacts on the destinations to be derived from the decrease of tourist demand, such as unemployment, social exclusion and political instability with danger to public safety as a result of breaks in the development of these regions (UNWTO 2008, pp. 67–68).

Weir (2017) argues that the integration of climate change predictions in tourism research is paramount to instruct the tourism strategic planning, emphasizing that the change in climate and tourism are not new, since they go back to the civilizations rise in the Indus Valley, Egypt, Mesopotamia and China, to the Dark Ages, to the Middle Ages' little Ice age and to the growth of modern mass tourism in the "Anthropocene".

Therefore, in light of climate change, Schott (2010) acknowledges the need to focus on different tourism consumers and adaptive innovation of tourism products (Schott 2010; Michailidou et al. 2016) by diversifying and promoting emergent and alternative tourism products, where eco-friendly and cultural tourism can play a major role.

As such, tourism should be managed not only contingently in a climate change framework, but also opportunistically, as the adaptation of tourism through diversification can lead to the sustainable development of the sector and regions that base their economic activity on it. So, in these cases, adaptation should be through tourism, but for communities, above all.

6.3 Touristic Regions, Climate Change and Green Infrastructure: A Contribute to Regional Resilience of Algarve Region, South of Portugal

In this subject, the Tourism of Portugal Institution has distinguished, pragmatically, two types of tourism products: Nature-based Tourism and Cultural and Landscape Touring. In the first one, the main tourist motivation is to live experiences of great symbolic value and to interact with and enjoy nature through sport activities and nature contemplation (THR 2006a, p. 9). The second is concerned with the discovery of the cultural and scenic heritage of the region through tours of routes or circuits (THR 2006b, p. 9). In this matter, landscape ecology and aesthetics are the basic essence of these touristic resources and the transformation into tourism products depends on its legibility (Queirós 2014, p. 177).

Rather than focus on products, the recent tourism strategy of Portugal—from now to 2027—is strategically focused on differentiating assets such as Climate and Sun Light, History and Culture, Nature and Water, where Landscape is diffused among them (TP 2017, pp. 46–48). It is based on a vision of "affirming tourism as a hub for economic, social and environmental development throughout the territory, positioning Portugal as one of the most competitive and sustainable tourism destinations in the world" (TP 2017, p. 52).

In this strategy, the major axes to be acknowledge here are (1) *Enhancing the Territory* with relevant lines of action, namely the conservation, enhancement and enjoyment of the natural and cultural heritage of Portugal, the promotion of urban regeneration of cities and regions, the sustainable development of touristic territories/destinations, and (2) the *Driving the Economy* axis, which assumes lines of action as to ensure the competitiveness of tourism companies in a short, medium and long-term perspective through the differentiation and innovation of regional tourism supply, to stimulate the circular economy in tourism by the consolidation of tourism in Portugal as reference in environmental, social, economic and governance aspects, and to affirm Portugal as an international benchmark in innovation, entrepreneurship and provision of tourism goods and services (TP 2017, p. 56).

In regards to climate change and tourism, a climate change impact assessment project carried out for Portugal (SIAM II) uses the temperature Bioclimatic Index Physiological Equivalent (Physiological Equivalent Temperature—PET), which is based on the indexation of the relationship between weather conditions and human physiological conditions (i.e. reactions like wearing more or less clothing and other human activities). The assessment reiterates that, in the projection to 2100, for the case study of Faro, the capital of the Algarve region district, there will be a reduction of thermal comfort in the summer months with an increasing frequency and duration of heat waves, which may result in an increase of diseases related to heat stress (Santos and Miranda 2006, p. 251) and vector-borne diseases. On other hand, there will be a gain of thermal comfort in April, May and October, as well as a decrease in the number of months and days with any degree of stress by cold, particularly in January (Santos and Miranda 2006, p. 251).

These changes will have a beneficial impact with respect to tourism in Algarve region, since they lead to the decrease of seasonality and increasing tourist exploitation of more pleasant months previously considered to part of the low season. So, it may result in an increment in tourist demand in spring and winter, particularly among the elderly population (Santos and Miranda 2006, p. 251).

However, beyond the general rise in temperature, the reduction of water availability and the potential decline in agricultural and forest productivity, as explained in the climate change impact assessment SIAM project (Santos et al. 2001, pp. 15–21), the Algarve region is especially vulnerable to the sea level rise. That said, the International Panel on Climate Change—IPCC (IPCC 2013, p. 1182; IPCC 2014b; APA 2014, p. 49) projects a mean sea level rise in the order of 0.42–0.98 m by 2100. It should be expected, as is happening in the Catalan coast (Jiménez et al. 2017) and in the Balearic Islands coast (Enríquez et al. 2017), the increase of coastal erosion phenomenon and a reduction or disappearance of sand beach areas (Ferreira et al. 2008, p. 15), a major touristic resource for *sun and beach* and the main engine of the regional development. There is a risk of drastic reduction in the touristic capacity of each beach, as a result, and the threat of damage to existing touristic-related buildings located near the coast (Ferreira et al. 2008, p. 15).

In the case of the Algarve coast, the Coastal Zone Management Plans (POOC) Burgau-Vilamoura and Vilamoura-Vila-Real-de-Santo-António already represent, respectively, 7.1% and 9.2% of the national budget allocated to coastal defence (APA 2014, p. 113).

On the other hand, the International Panel on Climate Change (IPCC 2014b, pp. 59–60) projects that the worsening of impacts in this century will lead to the inability of national states to finance, contingently and continuously, adaptation strategies related to climate change. This places doubts upon the future of coastal communities in their geographical locations, and also in their socio-economic dependence on conventional sun and beach mass tourism, as is the case in the Algarve region.

It is expected that there will be a substantial reduction in the tourist carrying capacity in beach areas, and tourists' respective unwillingness to visit Sun and Beach destinations. It is therefore important to present diversified and innovative alternatives to respond shifts on tourist behaviour (Gossling et al. 2012; Moreno 2010; UNWTO 2008), where green infrastructure can entail a transition path.

Green infrastructure can be defined as a network of natural and semi-natural spaces within, around and beyond the urban spaces, including gardens, lakes, parks, bike paths, green roofs, wetlands, greenways, river, streams, agricultural land and forest areas of sustainable use, which provides additional interconnection and enhanced resilience benefits (EEA 2011, pp. 30–35). This network is then based on the preservation and enhancement of the connectivity of ecosystems in order to maintain or gain the provision of ecosystem services, benefits and their resilience, which includes mitigation and adaptation to climate change.

Various academic and institutional forums have made important contributions to the understanding of (and policies surrounding) green infrastructure. For example, in 2013, the European Commission created the *Green Infrastructure Strategy—*

Enhancing Europe's Natural Capital (EC 2013, p. 2), which assuming it as a strategically planned network of natural and semi-natural areas, designed and managed to provide a wide range of ecosystem services. Incorporating green spaces (or blue, when integrating aquatic ecosystems), the greens infrastructures can be both rural and urban. Today, its importance is considered a 2020 priority and is reiterated by the EU Biodiversity Strategy to 2020 (EC 2013, p. 2).

Sussams et al. (2014, p. 186) report that the concept of a green infrastructure is congruently aligned with the *Ecosystem Approach*, developed in the Convention on Biological Diversity (CBD) COP5, which adopted a strategy for the integrated management of land, water and living resources, emphasizing the need for increased cooperation at all levels. The *Millennium Ecosystem Assessment* (MEA) brought into the sphere of economic science the importance of ecosystem services, presenting an opportunity recognition of the importance of planning and management of green infrastructures. The real bridge between this concept and adaptation to a changing climate was made by the *White Paper for Adaption to Climate Change*, which recognized a green infrastructure as being a crucial part in the provision of social and economic benefits in extreme weather conditions (Sussams et al. 2014, p. 186).

The main mitigation opportunities to be achieved by a green infrastructure are carbon sequestration (Nowak et al. 2013, p. 235), reduction of energy use for heating and/or cooling of buildings by increasing the area of green spaces, green roofs and green walls (Cit. Demuzere et al. 2014, p. 109), proximity production of agricultural and other goods (Beatley 2000, p. 7) and encouraging sustainable mobility through pedestrian and bike paths (NRDA 2010, p. 32).

The essential adaptation options based on a green infrastructure can go from the reduction of the urban heat island effect and increased thermal comfort (Oliveira et al. 2011, p. 2191), regulation of water quantity and quality (NRDA 2010, p. 36), storage and water drainage, reducing river flooding (Demuzere et al. 2014, p. 109), attenuation of sea flooding (NRDA 2010, p. 40), connectivity between habitats (EEA 2011, pp. 36–38; NRDA 2010, p. 46), benefits to public health, stimulating adaptability and education (Tzoulas et al. 2007, pp. 169–175; Demuzere et al. 2014, p. 111) to being an alternative opportunity for recreation and leisure, the centre point of this chapter (Samora-Arvela et al. 2016a, b).

In addition to climate challenges, the ageing population of the world at large and European Union (EU), specifically, is also an important consideration for the future of Mediterranean climate tourism. In 2010, for example, 17% of EU population was 65 years old or older, and that share is expected to move up to 30%. Despite the negative impact of recent economic and financial crisis on tourism in the EU, the 65 years old or older group contributed, substantially, to offset this negative impact. That is, between 2006 and 2011, the number of tourists fell in all age groups, except in the older age group, presenting a 10% increase over 2006 (EC 2010, pp. 5–6).

Tourism also shapes retirement migration. Some of the British retired persons living in Tuscany (Italy), Malta, the Costa del Sol (Spain) and in the Algarve region (Portugal), southern Europe, chose this region for spending their retirement due to prior experiences as tourists on these destinations of mass tourism, and by the expectation to find similar levels of comfort in their retirement as they did as tourists.

Thus, these migrants demand the facilities that only exist on these tourism destinations, because they are of mass tourism (Williams et al. 2000, p. 45).

Different is the approach of the tourists who were "explorers" or "individual mass tourists" willing to seek out areas that were less touched by mass tourism (Williams et al. 2000, p. 45).

That said, a strategy to strengthen regional resilience, that is "how a region or system responds to shock or disturbance and under these circumstances is able to ensure its continuous development" (Palekiene et al. 2015, p. 180) or the capacity of a region to sustain long-term development (Boschma 2015, p. 735; Christopherson et al. 2010) of Mediterranean mass tourism areas can pass, on one hand, by the containment of insistent and trend intent touristic-urban-coastal expansion and, on the other hand, by the consolidation other sustainable tourism products, namely Nature-based Tourism, and Cultural and Landscape Touring through the enhancement of endogenous resources (Monteiro et al. 2015) of green infrastructure. This strategic option expresses a cultural adaptation strategy (O'Brien and Hochachka 2009, p. 98) through spatial planning and integrated landscape management (Denier et al. 2015, p. 139), by the assumption that diversified regions have less risk of a shock being an enormous and negative impact on the local economy as a whole (Boschma 2015, p. 736; Christopherson et al. 2010).

6.4 Methodological Framework

The problem that has been presented requires a selection and analysis of several case studies of Mediterranean touristic destinations homologous in the exploration of the sun and beach touristic products. These case studies are in five geographic regions, namely the Rimini Province, Italy, the Antalya Province, Turkey, Cyprus, the Alicante Province, Spain, and the Algarve Region, Portugal (Fig. 6.2) and are examined in order to assess which (and to what extent) the various destinations, in each of their official tourism promotion websites, beside sun and beach, also recommend tourism practices in the inland areas. This is important in order to assess whether these provinces and regions develop greater resilience by tourism diversification—a potential shield to the indirect impacts of climate change. It is also important for developing an understanding and benchmarking of best practices of green infrastructure spatial planning, landscape management and tourism *praxis* for the Algarve region, Portugal.

6.4.1 Province of Rimini, Italy

The Riviera of Rimini is a famous destination for its five sun and beach seaside towns on the Adriatic Sea, and its well-renowned beaches of low water depth (Fig. 6.3).

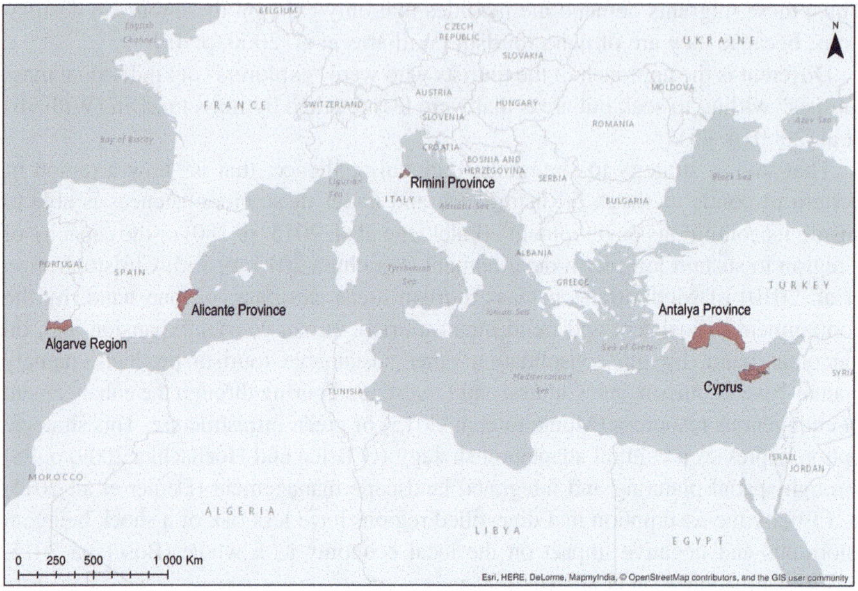

Fig. 6.2 Selected case studies

Fig. 6.3 Rimini beach (R/oddlysatisfying 2017)

Fig. 6.4 San Leo (Assessorato al Turismo 2017)

Here, the Assessorato al Turismo of the Comune di Rimini promotes, mainly and equally on its website (2017), the hinterland landscape—specifically the territory of Malatesta and the Montefeltro region, which are characterized by the contrast of beautiful valleys of the Conca and Marecchia rivers, punctuated by historical villages (Fig. 6.4). In these areas, tourists can find several natural parks, natural reserves and other places of natural interest.

6.4.2 *Antalya Province, Turkey*

According to the office of Turkey Tourism, the Gulf of Antalya, Province of Antalya, constitutes the *pearl* of Mediterranean and of Turkish Riviera (Turkey Tourism 2017). Antalya, a sea level rise vulnerable city, not only promotes its natural blue flag beaches, but also encourages swimming platforms over the sea as a measure of adaptation (Fig. 6.5) and sponsors natural attractions, such as Köprülü Canyon National Park (Fig. 6.6), Alara Ucansu Waterfall, Kursunlu Waterfall, Düden Waterfalls and Manavgat Waterfall, among many.

Fig. 6.5 Antalya beach

Fig. 6.6 Koprulu Canyon National Park (Turkey Tourism 2017)

6.4.3 Cyprus

Recognizing the compactness of its territory, Cyprus Tourism Organization (2017) presents, on their website, a clear and wide variety of touristic products and activities to do. These go beyond sun and beach, such as health and wellness, food and drink, weddings and honeymoons, culture and religion and nature. They also include activities based on cycling, climbing, nature trails, birdwatching and other sites of natural and cultural interest, interconnected by thematic routes.

Recently, ten national forest parks were established in Cyprus with the objective to support recreational opportunities in the hinterlands and to diversify tourism sector (Karanikola et al. 2017).

6.4.4 Alicante Province, Spain

This province is well known for its Costa Blanca tourist brand, where Benidorm is located—a dense and urbanized sun and beach destination. Beyond Costa Blanca brand, the official tourism portal of Spain also recommends visiting other tourist spots based on nature-based tourism or cultural-landscape touring, such as Elche Palm Grove, awarded as World Heritage by UNESCO, Castle of Santa Bárbara, Gothic Church of Orihuela, Terra Mítica theme park and Guadalest and Algar Fountains. In the inland of Alicante Province, there is the protected landscape of Serra del Maigmó y Serra del Sit, ideal for hiking and landscape-scenic touring of cultural heritage and panoramic views (Spain's official tourism portal 2017).

6.4.5 Algarve Region, Portugal

While Algarve Tourism Board has the main mission of improvement and qualification of the Algarve territory as a tourist destination, the Algarve Promotion Bureau is the non-profit making association in charge of the promotion of Algarve as a region to be visited. However, the Algarve Region promotion is heavily focused on sun and beach tourism, whereby these entities should follow the example of its congeners, especially Rimini Province, Antalya Province and Cyprus, in qualifying and promoting, effectively and integrally, the varied fullness of landscape units of Algarve region (Algarve Tourism Board and Algarve Promotion Bureau 2017).

Therefore, in terms of territory qualification and landscape promotion as a whole, it is possible to classify each of the studied Mediterranean destinations as following (Table 6.1).

Table 6.1 Degree of qualification and promotion of seaside and inland landscape in each tourist destination

Tourist destination	Degree of qualification and promotion of seaside and inland landscape
Rimini Province, Italy	Excellent
Antalya Province, Turkey	Excellent
Cyprus, Cyprus	Excellent
Alicante Province, Spain	Good, but there is still the need for improvement in promoting all the landscape units to be visited
Algarve region, Portugal	Good, but there is still the need for improvement in promoting all the landscape units to be visited

6.5 Algarve Region, Portugal: How to Plan a Transition?

In the framework of the Portuguese spatial planning, the Green Infrastructure Strategy is synonyms to the purpose of Ecological Structure, which includes the Public Hydric Domain (DPH), National Agricultural Reserve (RAN), National Ecological Reserve (REN) and the Fundamental Network for Nature Conservation (RFCN) (Magalhães 2013, pp. 6–9), where 2000 Natura Networks are integrated.

The Algarve region is characterized by three main landscape units (BCD of Fig. 6.2) (Cancela d'Abreu et al. 2004, pp. 171–218), where its diversity goes from *A—Monchique Mountains and Surroundings* (volcanic with *Eucalyptus globulus*, *Castanea sativa* and *Quercus* forest and shrublands); *B—Caldeirão Mountains* (abandoned Mediterranean xerophytic mountainous forest landscape); *C—Algarve Barrocal Midlands* (irrigated citrus orchards in limestone soils); *D—Western Coastal Landscape; E—Central Coastal Landscape* (dense touristic and habitational urban occupation); *F—Ria Formosa Wetlands* (natural park with high level of biodiversity and priority of conservation); *G—Vicentine Western Coastal Landscape* (coastal landscape with strong Atlantic influence); H—Sagres *and São Vicente Cape*; *I—Guadiana River Valley and Tributaries* (a close valley that constitutes the frontier between Portugal and Spain) and *J—Guadiana River Mouth* (Fig. 6.7).

In 1967, the landscape architect António Viana Barreto and his team prepared the study *Landscape Planning of Algarve Region,* which aimed a non-disruptive territorial development in conjugation with landscape endogenous resources preservation and enhancement (Viana-Barreto et al. 1971, pp. 123–127). However, despite the considerable efforts of this plan, this goal was not followed, and the major regional driver was and has been the expansionist boom of urban fabric along the coastal zone. This was registered by Botequilha-Leitão et al. (2011, p. 3) in the analysis of changes between a cartographic series on various land uses of the Algarve region (Corine Land Cover 1985–2006 and Land Use Occupation Map—COS—1990–2007), concluding that 35% of the area within the 2 km area of distance from the coast is urbanized.

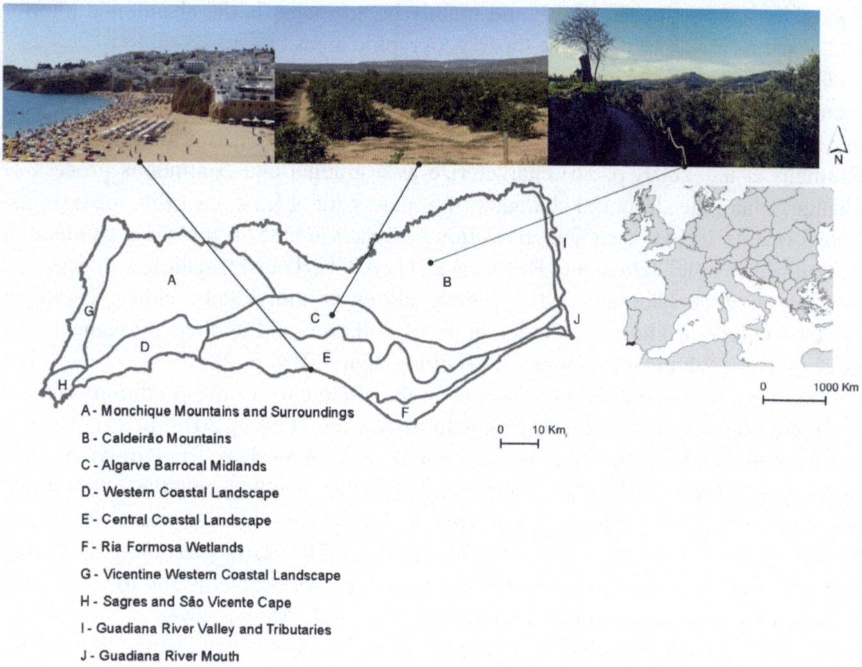

Fig. 6.7 Algarve region landscape units (Cancela d'Abreu et al. 2004, pp. 171–218)

This inordinate and huge coastal sprawl is derived from the promotion of unqualified sun and beach tourism, and the worsening of economic and social inequalities between coastal zones and the most interior landscape units (mountains and midlands) (CCDRAlg 2006, p. 6).

It should be noted that 39.6% of the Algarve region population is employed in *trade, touristic accommodation, transport and communication* and *financial activities, real estate and services to companies* represents 12.7% of employment (Tourism of Portugal Institution 2014, p. 42).

Thus, the current Regional Strategic Plan for the Algarve Region (CCDRAlg 2006) and the newly established Operational Programme for Algarve Region for 2014–2020 (CCDRAlg 2016a) assume the urgency of combating social disparities through the integration of value in low-density territories the diversification of tourism products within the region (CCDRAlg 2016b, pp. 14–16).

Simultaneously, the present and future climate change and respective sea level rise through the potential reduction of beach area can be disruptive to a region, where the main economic activity is concentrated, profoundly, on the coast.

In this way, the regions whose socio-economic decisions are exclusively depended on sun and beach tourism products may face a *Dantesque* scenario for which they are not, minimally, prepared. As such, several strategies could be considered, including the increase of diversification in tourism products through the promotion of nature-based tourism and cultural-landscape touring, which can take place near the coast, but should mainly be promoted in the abandoned interior landscape as a way to revitalize those geographic areas.

From this comes the urgency of considering alternative visions (e.g. the enhancement of a green infrastructure) to strengthen regional resilience for tourism across various landscape units in this region. That said, a transition is a long process that Rotmans et al. (2001, p. 16) characterize as a gradual and continuous process of change, where the structural character of a society (or at least one of its subsystems) transforms itself. In this subject, transition pathways are the "actual ways a transition can unfold in time" (Haan and Rotmans 2011, p. 93). Transformation is understood here as adaptive alternatives for "organizations or individuals, either forced by systems failure or chosen in anticipation of collapse, and as their movement to a novel social-ecological system state" (Pelling et al. 2014, p. 2).

So, in order to avoid the late defence to the emergence of the conditions created by future impacts and its catastrophic natural disasters (Pelling 2011, p. 95), Algarve region policymakers and inhabitants should free themselves from trend or path dependence, proactively and strategically, creating a transformative or, at least, teleological path of reinforcing resilience and sustainable development (Haan and Rotmans 2011, pp. 99–100; Garud et al. 2010, pp. 768–770), presenting appealing and imaginative visions to overcome the persistence of problems and to inspire the support of a wide range of different actors (Voβ et al. 2009, p. 284).

Thus, an important question is whether or not is possible to change the current practice of tourism in the Algarve region through the sustainable management of a green infrastructure. Green infrastructure can contribute to the resilience of a tourism-based region, helping to adapt to the negative impacts of climate change. How can policy reinforce green infrastructure? The following table (Table 6.2) elucidates on this matter.

So, it's paramount to determine a transition pathway to nature-based tourism and cultural-landscape touring, which can be based on the satisfaction of complementary preferences of tourists and residents for activities, beyond beach recreation, such as bird watching, canoeing, biking, swimming in natural pools (Fig. 6.8), accommodation in rural areas and gastronomy tasting, among others.

In Portugal, green infrastructures are already mapped at national and regional level. The greatest aspiration should therefore be its delineation at the local level through the design of *Municipal Ecologic Structures* at the revision of *Master Plans* (TP 2016, pp. 35–36). In the case of the Algarve region municipalities, they should not strictly focus on the building prohibition, but identify, sustainable, recreation, leisure and touristic spaces based on nature, culture and landscape. This is a way to diversify tourism and to create alternatives to the sea level rise highly vulnerable sun and beach cluster. Public initiatives for planning these municipal green infrastructures must go beyond mapping, thus, creating incentive mechanisms, education programs and helping private stakeholders in their concerns and sustainable investment intents, a regnant condition for public policy success (Samora-Arvela et al. 2017).

Thereby, policymakers and tour operators could contribute, proactively, to the planning process of a green infrastructure in light of this sustainable touristic purpose

Table 6.2 Transition pathway of regional resilience enhancement of Mediterranean destinations through the communion of tourism diversification and green infrastructure implementation

<table>
<tr><td colspan="3">Strategic axis</td></tr>
<tr><td colspan="3">Diversifying Mediterranean tourism under climate change challenges as a strategy of regional resilience strengthening for Algarve region, south of Portugal, through green infrastructure enhancement</td></tr>
<tr><td colspan="3">Goals</td></tr>
<tr><td colspan="3">Diversification of tourism, recreation and leisure activities: identification of areas of sustainable use for tourists and inhabitants
Pedestrian and cycling mobility: delineation of tracks
Conservation: of ecologically sensitive areas and its connectivity
Preserving Mediterranean landscape mosaic: regional products
Preservation of cultural heritage: itineraries and accommodation in restored vernacular constructions</td></tr>
<tr><td>Stakeholders</td><td colspan="2">Lines of action</td></tr>
<tr><td>Tourism operators and touristic accommodation companies</td><td colspan="2">Promotion of routes, tours and accommodation in rural areas in close connection with the local community
Use of regional products in tourism activity and its promotion (food and handicraft) through labels of origin, selling or offering at the accommodation establishments
Investment in green roofs and green walls as a factor of adaptive and aesthetic rehabilitation of touristic buildings with low cultural or architectural value
Provision of funds</td></tr>
<tr><td rowspan="3">Authorities in spatial planning and tourism planning</td><td>European Union</td><td>Support for green infrastructure through research and demonstration
Provision of funds</td></tr>
<tr><td>National state</td><td>Budget administration and submitting applications
Application of fiscal incentives and revolving loans to tourism operators, touristic accommodation companies and landowners/users who invest in green infrastructure and tourism projects, including urban rehabilitation
Provision of funds</td></tr>
<tr><td>Regional and local authorities</td><td>Supervisory, coordination and management role in implementation
Development of educational materials, communication campaigns and capacity building activities for stakeholders
Development and implementation of a visitor guidance concept, supporting sustainable tourism
Regulation of vernacular architecture preservation
Monitoring activities</td></tr>
<tr><td>Scientific and technical experts</td><td colspan="2">Advisory role during planning and implementation stages
Serving as specialists for specific theme or area of the project, supporting decision-making processes</td></tr>
</table>

(continued)

Table 6.2 (continued)

Landowners/users, and local community	Investment in low-density touristic accommodation in rural space working with the tourism operators Collaboration in the planning and the making of routes and tours with tourism operators Production of regional products with high quality Preservation of vernacular architecture and landscape
Tourists	Creative tourism to deepen contact with the local culture by directly participating in cultural/creative activities and being involved in the creative life of the destination, choosing nature-based tourism activities and cultural-landscape touring
Inhabitants	Beyond the inclination for sun and sea, it's necessary to promote the landscape value and culture of the hinterlands

Adapted from Naumann et al. 2011, pp. 34–40

Fig. 6.8 Natural pool in Santa Luzia, Tavira municipality, Algarve region

in order to galvanize regional competitiveness as a *modus* to adapt this region to future challenges.

6.6 Conclusions

Among the several impacts expected on Mediterranean climate tourism, the sea level rise and the potential beach area reduction could put a great impediment to well-being and development of tourism-based regions.

One alternative is to strategically study a transition of tourism towards a sustainable approach, ensuring the regional resilience of tourism destination regions. Green infrastructure can play a major role in presenting alternatives to sun and beach, creating sustainable tourism products, diversifying activities, and strengthening regional resilience through the enhancement of nature-based tourism and cultural-landscape touring. In order to do so, it is paramount for all stakeholders to work together to build a future, where tourism, as the main socio-economic activity of the Mediterranean coastal areas, transits towards resiliency and sustainability. Furthermore, under the potential occurrence of sand beach area reduction, it is necessary to research what complementary (and substitutive preferences) tourists and inhabitants have beyond the inclination for sun and beach. Encouraging the enjoyment of other activities, such as nature-based tourism activities and cultural-landscape touring of the Mediterranean regions, will ultimately assist the sustainable urban and rural spatial planning.

The multifunctionality of a green infrastructure appears to be a defence to climate change degradations, but it should be noted that its contribution is not unlimited. As such, only a social transformation, a consequent "biophilic conscience" (Beatley 2000) and the recognition of the "circular metabolism of human systems" (Pearce and Turner 1990, pp. 57–89; Boulding 1997, pp. 3–14) can solve the causes of climate change, while green infrastructure can, only and to some extent, attenuate its impacts.

Nevertheless, tourism is not wholly innocent, since it contributes to the cause (s) of climate change (Prista 2015; Samora-Arvela et al. 2016a, b). Re-examining assumptions about the impacts of tourism and its role in climate change, particularly for coastal Mediterranean regions like the Algarve, provides an opportunity to focus planning efforts (public and private) on regional green infrastructure in order to ensure the sustainable development of this region, overall. While Herculean, this strategy could help coastal communities manage the impacts of climate change.

Acknowledgement This study was financed by FCT—Foundation for Science and Technology through the PhD grant SFRH/BD/102328/2014. The authors thank the support given by ICS—Social Science Institute, University of Lisbon, CICS.NOVA—Interdisciplinary Centre of Social Sciences, Faculty of Social Sciences and Humanities, New University of Lisbon, Research Centre for Spatial and Organizational Dynamics, University of Algarve, and Laboratory for Geocomputation, Department of Geography and Environmental Studies, Faculty of Arts, Ryerson University.

References

Algarve Tourism Board and Algarve Promotion Bureau. (2017). Algarve tourism. Retrieved from http://www.visitalgarve.pt/visitalgarve/vPT/VivaOAlgarve/

Amelung & Moreno. (2009). *Impacts of climate change on tourism in Europe*. Seville: European Commission, Institute for Prospective Technological Studies.

APA. (2014). *Gestão da Zona Costeira – O Desafio da Mudança – Relatório do Grupo de Trabalho do Litoral*. Lisbon: Environment Portuguese Agency.

Assessorato al Turismo – Comune di Rimini. (2017). *Rimini tourism*. Retrieved from http://www.riminiturismo.it/en/visitors/enjoy-and-relax/beach

Beatley, T. (2000). *Green urbanism: Learning from European cities*. Washington: Island Press.

Becken, S., & Hay, J. (2007). *Tourism and climate change – Risks and opportunities*. Ontario: Channel View Publications.

Biesbroek, G. R., Swart, R. J., & Van der Knaap, W. G. M. (2009). The mitigation–adaptation dichotomy and the role of spatial planning. *Habitat International, 33*, 230–237. https://doi.org/10.1016/j.habitatint.2008.10.001

Boschma, R. (2015). Towards an evolutionary perspective on regional resilience. *Regional Studies, 49*(5), 733–751. https://doi.org/10.1080/00343404.2014.959481

Botequilha-Leitão, A. (2011). *Landscape changes in the Algarve region, Portugal ('85-'07) – diagnosis, prospective and a proposal for a green infrastructure in the Algarve Coast*. International Conference "Green Infrastructure for Biodiversity" Forum Biodiversity. Estoril: Congress Center of Estoril.

Boulding, K. (1997). The economics of the coming spaceship earth. In R. Robert Costanza, C. Perrings, & C. J. Cleveland (Eds.), *The development of ecological economics* (pp. 3–14). Northampton: Edward Elgar Publishing.

Bowman, K. S. (2011). Sustainable tourism certification and state capacity: Keep it local, simple, and fuzzy. *International Journal of Culture, Tourism and Hospitality Research, 5*, 269–281. Retrieved from http://www.emeraldinsight.com/doi/abs/10.1108/17506181111156961

Bramwell, B., & Lane, B. (2012). Towards innovation in sustainable tourism research. *Journal of Sustainable Tourism, 20(1)*, 1–7. https://doi.org/10.1080/09669582.2011.641559

Cancela d'Abreu, A., Pinto Correia, T., & Oliveira, R. (2004). *Contributos para a Identificação e Caracterização da Paisagem em Portugal Continental* (Vol. V, pp. 171–218). Lisbon: Direcção-Geral do Ordenamento do Território e Desenvolvimento Urbano.

Casimiro, E., Gomes, A., & Almeida, S. (2010). *Climate change strategic plan of Cascais municipality*. Lisbon: University of Lisbon Faculty of Sciences.

CCDRAlg. (2006). *Plano Regional de Ordenamento do Território do Algarve*. Faro: Comissão de Coordenação e Desenvolvimento da Região do Algarve.

CCDRAlg. (2016a). *Programa Operacional ao Abrigo do Objectivo de Investimento no Crescimento e no Emprego*. Faro: Comissão de Coordenação e Desenvolvimento da Região do Algarve.

CCDRAlg. (2016b). Algarve – Preparar o Futuro – Diagnóstico Prospectivo 2014 – 2020. Faro: Comissão de Coordenação e Desenvolvimento da Região do Algarve.

Ceballos-Lascuráin, H. (1988). The future of "ecotourism". *Mexico Journal, 27*, 13–14.

Christopherson, S., Michie, J., & Tylerc, P. (2010). Regional resilience: Theoretical and empirical perspectives. *Cambridge Journal of Regions, Economy and Society, 3*, 3–10. Retrieved from https://doi.org/10.1093/cjres/rsq004

Cyprus Tourism Organisation. (2017). *The official portal of Cyprus tourism*. Retrieved from http://www.visitcyprus.com/index.php/en/

Demuzere, D., Orru, K., Heidrich, O., Olazabal, E., Geneletti, D., Orru, H., Bhave, A. G., Mittal, N., Feliu, E., & Faehnle, M. (2014). Mitigating and adapting to climate change: Multifunctional and multiscale assessment of green infrastructure. *Journal of Environmental Management, 146*, 107–115. Retrieved from https://doi.org/10.1016/j.jenvman.2014.07.025

Denier, L., Scherr, S., Shames, S., Chatterton, P., Hovani, L., & Stam, N. (2015). *The little sustainable landscapes book*. Oxford: Global Canopy Programme.

Dernoi, L. (1981). Alternative tourism: Towards a new style in north-south relations. *Tourism Management, 81*, 253–264.

EC. (2010). *Europe, the best destination for seniors*. Brussels: European Commission.

EC. (2013). *Building a green infrastructure for Europe*. Brussels: European Commission.

EEA. (2011). *Green infrastructure and territorial cohesion. The concept of green infrastructure and its integration into policies using monitoring systems* (EEA technical report no. 18/2011). Copenhagen: European Environment Agency.

Enríquez, A., Marcos, M., Álvarez-Ellacuría, A., Orfila, A., & Gomis, D. (2017). Changes in beach shoreline due to sea level rise and waves under climate change scenarios: Application to the Balearic Islands (western Mediterranean). *Natural Hazards and Earth System Sciences, 17*, 1075–1089. Retrieved from https://doi.org/10.5194/nhess-17-1075-2017

Fennell, D. (1999). *Ecotourism: An introduction*. New York: Routledge.

Ferreira, Ó., Dias, J. A., & Taborda, R. (2008). Implications of sea-level rise for continental Portugal. *Journal of Coastal Research, 24*(2), 317–324. ISSN: 0749-0208. https://doi.org/10.2112/07A-0006.1

Garud, R., Kumaraswamy, A., & Karnoe, P. (2010). Path dependence or path creation? *Journal of Management Studies, 47*(3), 760–774. https://doi.org/10.1111/j.1467-6486.2009.00914.x

Gossling, S., Scott, D., Hall, C. M., Ceron, J., & Dubois, G. (2012). Consumer behaviour and demand response of tourists to climate change. *Annals of Tourism Research, 39*(1), 36–58. https://doi.org/10.1016/j.annals.2011.11.002

Haan, J., & Rotmans, J. (2011). Patterns in transitions: Understanding complex chains of change. *Technological Forecasting and Social Change, 78*, 90–102. Retrieved from https://doi.org/10.1016/j.techfore.2010.10.008

IPCC. (2013). *AR5 – Climate Change 2013 – The physical science basis. Contribution of Working Group I to the Fifth Assessment Report of the Intergovernmental Panel on Climate Change*. Cambridge: Cambridge University Press.

IPCC. (2014a). *AR5 – climate change 2014 – Impacts, adaptation, and vulnerability*. Part B: Global and Sectoral Aspects. Contribution of Working Group II to the Fifth Assessment Report of the Intergovernmental Panel on Climate Change. Cambridge: Cambridge University Press.

IPCC. (2014b). *AR5 – climate change 2014 – Impacts, adaptation, and vulnerability*. Part A: Global and Sectoral Aspects. Contribution of Working Group II to the Fifth Assessment Report of the Intergovernmental Panel on Climate Change. Cambridge: Cambridge University Press.

Jiménez, J. A., Valdemoro, H., Bosom, E., Sánchez-Arcilla, A., & Nicholls, R. (2017). Impacts of sea-level rise-induced erosion on the Catalan coast. *Regional Environmental Change, 17*, 593–603. Retrieved from https://doi.org/10.1007/s10113-016-1052-x

Karanikola, P., Thomas Panagopoulos, T., & Tampakis, S. (2017). Weekend visitors' views and perceptions at an urban national forest park of Cyprus during summertime. *Journal of Outdoor Recreation and Tourism, 17*, 112–121. Retrieved from https://doi.org/10.1016/j.jort.2016.10.002

Kourtit, K. (2017). Smart development, spatial sustainability and environmental quality. *Habitat International, 68*, 1–2. https://doi.org/10.1016/j.habitatint.2017.08.001

Liu, Z. (2003). Sustainable tourism development: A critique. *Journal of Sustainable Tourism, 11*, 459–475. https://doi.org/10.1080/09669580308667216

Magalhães, M. R. (2013). *Estrutura Ecológica Nacional – Proposta de Delimitação e Regulamentação*. Lisbon: Centro de Estudos de Arquitectura Paisagista "Professor Caldeira Cabral", ISA Press.

Michailidou, A. V., Vlachokostas, C., & Moussiopoulos, N. (2016). Interactions between climate change and the tourism sector: Multiple-criteria decision analysis to assess mitigation and adaptation options in tourism areas. *Tourism Management, 55*, 1–12. https://doi.org/10.1016/j.tourman.2016.01.010

Monteiro, V., Painho, M., & Vaz, E. (2015). Is the heritage really important? A theoretical framework for heritage reputation using citizen sensing. *Habitat International, 45*, 156–162. https://doi.org/10.1016/j.habitatint.2014.06.022

Moreno, A. (2010). Mediterranean tourism and climate (change): A survey-based study. *Tourism Planning and Development, 7*, 253–265.

Naumann, S., McKenna, D., Kaphengst, T., Pieterse, M., & Ryament, M. (2011). *Design, implementation and cost elements of green infrastructure projects*. Final report to the European Commission, DG Environment, Contract no. 070307/2010/577182/ETU/F.1, Ecologic Institute and GHK Consulting.

Nowak, D. J., Greenfield, E., Hoehn, R., & Lapoint, E. (2013). Carbon storage and sequestration by trees in urban and community. *Environmental Pollution, 178*, 229–236. Retrieved from https://doi.org/10.1016/j.envpol.2013.03.019

NRDA. (2010). *Green infrastructure: How and where can it help the northwest mitigate and adapt to climate change?* Warrington: Northwest Regional Development Agency.

O'Brien, K., & Hochachka, G. (2009). Integral adaptation to climate change. *Journal of Integral Theory and Practice, 5*(1), 89–102. Retrieved from https://integralwithoutborders.net/sites/default/files/resources/O%27Brien_Hochachka_FINAL.pdf

Oliveira, S., Andrade, H., & Vaz, T. (2011). The cooling effect of green spaces as a contribution to the mitigation of urban heat: A case study of Lisbon. *Building and Environment, 46*, 2186–2194. Retrieved from https://doi.org/10.1016/j.buildenv.2011.04.034

Palekiene, O., Simanaviciene, Z., & Bruneckiene, J. (2015). The application of resilience concept in the regional development context. *Social and Behavioral Sciences, 213*, 179–184. Retrieved from https://doi.org/10.1016/j.sbspro.2015.11.423

Pearce, D. W., & Turner, K. (1990). *Economics of natural resources and the environment*. London: Harvester Wheatsheaf.

Pelling, M. (2011). *Adaptation to climate change*. New York: Routledge.

Pelling, M., O'Brien, K., & Matyas, D. (2014). Adaptation and transformation. In J. Birkmann & R. Mechler (Eds.), *Advancing climate change and risk management*. London: King's College London.

Prista, P. (2015, November/December 9). *Turismo e Alterações Climáticas*. Smart cities. Retrieved from http://smart-cities.pt/pt/noticia/turismo-e-alteracoes-climaticas1612

Queirós, A. S. (2014). Turismo de Paisagem. In A. Veríssimo-Serrão (Ed.), *Filosofia e Arquitectura da Paisagem, Um Manual* (pp. 177–187). Lisbon: University of Lisbon, Philosophy Centre.

R/oddlysatisfying. (2017). *R/oddlysatisfying*. Retrieved from https://www.reddit.com/r/oddlysatisfying/comments/6bbqpk/parasols_rimini_beach_italy/

Romão, J., & Neuts, B. (2017). Territorial capital, smart tourism specialization and sustainable regional development: Experiences from Europe. *Habitat International, 68*, 64–74. https://doi.org/10.1016/j.habitatint.2017.04.006

Rotmans, J., Kemp, R., & van Asselt, M. (2001). More evolution than revolution: Transition management in public policy. *Foresight, 3*(1), 15–31. Retrieved from https://doi.org/10.1108/14636680110803003

Samora-Arvela, A., Ferrão, J. Ferreira, J., Oliveira, R., Panagopoulos, T., Vaz, E. (2016a, November 7–9). Infraestrutura Verde e Alterações Climáticas: um contributo para a mitigação e adaptação do território às alterações climáticas. In R. García Marín, F. Alonso Sarría, F. Belmonte Serrato, & D. Moreno Muñoz (Eds.) X*V Coloquio Ibérico de Geografía. Retos y tendencias de la Geografía Ibérica*, 376–381, Murcia: Universidad de Murcia-AGE-APG. ISBN: 978-84-944193-4-4

Samora-Arvela, A., Ferreira, J., Panagopoulos, T., & Vaz, E. (2016b, May 5–6). Tourism and climate change: Mitigation and adaptation in the path of transformation. In: *Proceedings of the international congress on interdisciplinarity in social and human sciences*. Faro: University of Algarve.

Samora-Arvela, A., Ferrão, J., Ferreira, J., Panagopoulos, T., & Vaz, E. (2017). *Green infrastructure, climate change and spatial planning: Learning lessons across borders*. II international congress of interdisciplinarity in social and human sciences. Faro: University of Algarve, Faculty of Economics, Research Centre for Spatial and Organizational Dynamics.

Santos, F. D., & Miranda, P. (2006). *Alterações Climáticas em Portugal – Cenários, Impacts e Medidas de Adaptação – Projeto SIAM II*. Lisboa, Gradiva.

Santos, F. D., Forbes, K., & Moita, R. (2001). *Mudança Climática em Portugal. Cenários, Impacts e Medidas de Adaptação – SIAM – Sumário Executivo e Conclusões*. Lisboa: Gradiva.

Schott, C. (2010). Tourism and the implications of climate change: Issues and actions. *Bridging Tourism Theory and Practice, 3*, 319–362. Retrieved from http://eds.b.ebscohost.com/eds/

ebookviewer/ebook/ZTAwMHR3d19fMzUxNzQ3X19BTg2?sid=8ecb1d2f-0733-4fee-b04b-c458d3297f6e@pdc-v-sessmgr01&vid=3&format=EB&rid=2

Scott, D. (2011). Why sustainable tourism must address climate change. *Journal of Sustainable Tourism, 19(1)*, 17–34. https://doi.org/10.1080/09669582.2010.539694

Sharpley, R. (2000). Tourism and sustainable development: Exploring the theoretical divide. *Journal of Sustainable Tourism, 8(1)*, 1–19. https://doi.org/10.1080/09669580008667346

Silva, L. (2009). *Casos no Campo – Etnografia do Turismo Rural em Portugal*. Lisbon: University of Lisbon, Social Science Institute, Imprensa de Ciências Sociais.

Simpson, M. C., Gossling, S., Scott, D., Hall, C. M., & Gladin, E. (2008). *Climate change and adaptation and mitigation in tourism sector: Frameworks, tools and practices*. Paris: UNEP, University of Oxford, UNWTO, WMO.

Spain's official tourism portal. (2017). *Province of Alicante*. Retrieved from http://www.spain.info/en/que-quieres/ciudades-pueblos/provincias/alicante__alacant.html

Sussams, L. W., Sheate, W. R., & Eales, R. P. (2014). Green infrastructure as climate change adaptation policy intervention: Muddying the waters or clearing a path to a more secure future? *Journal of Environmental Management, 147*, 184–193. Retrieved from https://doi.org/10.1016/j.jenvman.2014.09.003

THR – Asesores en Turismo Hotelería y Recreación, S.A. (2006a). *Nature-based tourism*. Lisbon: Tourism of Portugal.

THR – Asesores en Turismo Hotelería y Recreación, S.A. (2006b). *Cultural and landscape tourism*. Lisbon: Tourism of Portugal.

TP. (2016). *Tourism sector approach in the master plan review – Guidance Guide*. Lisbon: Tourism of Portugal.

TP. (2017). *Tourism strategy 2027*. Lisbon: Tourism of Portugal.

Turismo de Portugal. (2014). *Plano de Marketing Estratégico para o Turismo do Algarve*. Faro: Turismo de Portugal – Turismo do Algarve.

Turkey Tourism. (2017). *Go turkey tourism*. Retrieved from http://www.goturkeytourism.com/destinations-turkey/antalya-city-in-turkey.html#

Tzoulas, K., Korpela, K., Venn, S., Yli-Pelkonen, V., Kazmierczak, A., Niemela, J., & James, P. (2007). Promoting ecosystem and human health in urban áreas using green infrastructure: A literature review. *Landscape and Urban Planning, 81*, 167–178. Retrieved from https://doi.org/10.1016/j.landurbplan.2007.02.001

UNWTO. (2008). *Climate change and tourism – Responding to global changes* (pp. 61–173). Madrid: United Nations World Tourism Organization.

Viana-Barreto, A., Dentinho, A., & Castelo-Branco, A. (1971). Ordenamento Paisagístico do Algarve. Arquitectura, n° 121–122, 4ª série, 123–127.

Voβ, J., Smith, A., & Grin, J. (2009). Designing long-term policy: Rethinking transition management. *Policy Sciences and Business Media, 42*, 275–302. https://doi.org/10.1007/s11077-009-9103-5

Weaver, D. (2006). *Sustainable tourism – Theory and practice*. Oxford: Elsevier.

Weir, B. (2017). Climate change and tourism – Are we forgetting lessons from the past? *Journal of Hospitality and Tourism Management, 32*, 108–114. https://doi.org/10.1016/j.jhtm.2017.05.002

Williams, A., King, R., Warnes, A., & Patterson, G. (2000). Tourism and International retirement migration: New forms of old relationship in southern Europe. *Tourism Geographies, 2*(1), 29–49. https://doi.org/10.1080/146166800363439

Wood, K., & House, S. (1991). *The good tourist: A worldwide guide to the green traveller*. New York: Mandarin.

Zolfani, S. H., Sedaghat, M., Reza Maknoon, R., & Zavadskas, E. K. (2015). Sustainable tourism: A comprehensive literature review on frameworks and applications. *Economic Research-Ekonomska Istraživanja, 28(1)*, 1–30. https://doi.org/10.1080/1331677X.2014.995895

Part III
Towards Strategies for Resilience

Chapter 7
Design Solutions for Resilience

Davide Fassi and Carla Sedini

7.1 Introduction: How Design Is Dealing with Resilience

Resilience is identified as the capacity of communities and institutions to face and react to environmental, economic and social problems in effective and innovative ways. Because of its own nature, design and design research is able to create and operate as an activator for the right conditions to engage resilient processes (Walker et al. 2004; Colucci 2012; Graziano 2012; Pisano 2004; Rodin 2014; Pinto 2015).

In the literature on economics, resilience has mainly to do with the ability of industries to adapt their strategies to answer to the economic changes and crisis as soon as they appear (Christopherson et al. 2010). Stephane Hallegatte (2014) developed the notion of economic resilience, according to which the combination of microeconomic and macroeconomic resilience helps to avoid (or diminish) damage to the welfare system.

Both social and political factors have a dramatic influence on the resilient capacities of a geographical area. Creative and cultural industries seem to have the highest levels of resilience on the crisis, however, even this sector was penalized, particularly the younger generations of workers (Stumpo and Manchin 2014). From an economic point of view, regional resilience depends also on the capacity of enterprises for innovation; the ability of the entrepreneurial environment to create new opportunities; and the attitude of institutions and individuals to be reactive (Sabatino 2015). As Sotarauta (2005) explains, policies aimed at attracting creative and innovative knowledge skills, based on the collaboration between academia, government and enterprises (Etzkowitz and Leydesdorff 1997, 2000), can be crucial for resilience.

D. Fassi · C. Sedini (✉)
Design Department, Politecnico di Milano, Milano, Italy
e-mail: davide.fassi@polimi.it

H. Pinto et al. (eds.), *Resilience and Regional Dynamics*, Advances in Spatial Science, https://doi.org/10.1007/978-3-319-95135-5_7

However, on a small scale, resilience is concerned with the maintenance and improvement of the quality of life of individuals, which can be achieved thanks to the creation of desirable contextual conditions. Many intangible elements, which have to do more with social capital than with economic capital, have to be taken into consideration and design methods can be very useful in highlighting them and making them explicit. Therefore, narrative, participatory, and co-design approaches can create and facilitate visible connections, which on a small scale can help resilience (Fassi and Sedini 2017).

According to Manzini (2015), resilience must be regarded with a positive meaning by moving from a mainly defensive one to a more positive, which sees resilience as a human (positive) characteristic. Therefore, designing for resilience involves taking into account four different features of the socio-technical system: diversity, efficiency, adaptability, and cohesion (Fiksel 2003), which we will address in the following paragraphs.

The focus of this chapter will mainly be on the methods that are typical of design processes and which can be systematized in order to become strategies of resilience. Our attention will be on the area of Milan, in the Lombardy region of Italy. In terms of resilience, Milan already takes part in the Rockefeller Foundation project *100 Resilient Cities*[1] focusing on suburban areas, climate problems and housing issues. In addition to this project, Fondazione Cariplo, a philanthropic organization, launched a funding program in 2015 called *Comunità Resilienti* ("Resilient Communities") in order to encourage and contribute to the propagation of initiatives, which, being based in environmental issues, were able to promote the notion of whole community resilience. In this chapter we are going to frame the "state of the art" in terms of the resilience of this geographical area.

The last section of this chapter will be dedicated to three case studies carried out by the Department of Design of the Politecnico di Milano that illustrate the relationship between design practices and resilience, which include: Coltivando, CCAlps-Creative Companies in Alpine Space, and ARNOLD.

7.2 Design Tools and Methods

"Design tools" facilitate the success of projects that have the involvement of institutions and citizens as a principal goal. Designing solutions for resilience means expanding the idea of design from an exclusive act done by designers to something that allows people to be creative themselves and to solve unexpected changes in their everyday lives. Throughout the years, designers have moved from being solution-developers *for* people to professionals creating *with* people, thus allowing people to design by and for themselves (Brown 2009). At a larger scale,

[1]Collaboration between the Municipality of Milan, the Agency for Mobility Environment and Territory, Politecnico di Milano and Kyoto Club. http://www.100resilientcities.org/

design can be used to develop innovative policies, to connect institutions and citizens, and to have a multidisciplinary approach to generating new economic and social value (European Commission, Implementing an Action Plan for Design-Driven Innovation, 2013).

Tools and methods in design are more common in the participatory practices that are used and tested in developing design outputs from emerging opportunities. These practices could be classified as follows (Björgvinsson, Ehn, and Hillgren 2010; Ehn 2010; Manzini 2015):

- Highly dynamic processes, including linear co-design processes and consensus-building methodologies (i.e. the most traditional view of participatory design);
- Creative and proactive activities—where the role of a design expert includes that of mediator (between different interests) and facilitator (of other participants' ideas and initiatives), while also including their individual creativity and culture;
- Complex co-design activities that call for prototypes, mock-ups, design games, models, sketches and other materials in order to be promoted, sustained and oriented. This involves a set of dedicated and designed artefacts that are the design experts' responsibility to conceive and create.

Research that has been developed by the POLIMI Desis Lab[2] in the Design Department of the Politecnico di Milano will be presented in this chapter, taking into account these three ways of developing solutions for design opportunities. These research activities put design strategies in place in order develop business ideas and to improve of the quality of life of a neighbourhood.

In 2010, the Polimi DESIS Lab began a process of creating a Living Lab on the university campus located in the district of Bovisa. This was done to allow people in the neighbourhood to both discover and use it, and to let the students come into contact with real life problems and opportunities. Research activities were thus developed through educational ones—a kind of action research that engaged students in parts of the projects, asking them to generate ideas that contributed to the work overall. It is documented that, through hands-on, project-based learning, experiential education facilitates student success (Cantor 1995). By making research through design (Frayling 1993), a continuous process of implementation of the results is allowed. The prototyping of the solutions is the core of the research led by the team because making the ideas quickly tangible, by having the people at the centre of them, allows for exploration of many ideas in parallel (Brown 2009). At the same time, the role of the designer in these processes of engagement with people and development of social innovations has to be understood in terms of timing in order to make initiatives self-sufficient and the community 'competent' (Meroni and Sangiorgi 2011). This approach lets the team focus its research on the public space as a complex system to be investigated in its embodied features of resilience through the use of design tools and methods.

---

[2]www.desis-network.org

A livable city is a city for everyone (Hahlweg 1997), where common spaces are the centres of social life (Salzano 1997). According to the research team, a more hospitable and livable city could even begin to emerge from small actions in public spaces. The design actions developed in the past few years were mostly inclusive and related primarily to the two communities—(1) the university, including students, faculty and staff, and (2) the external one, including both the residents of the Bovisa neighbourhood and the city of Milan, at large.

This was achieved through a series of initiatives such as events, classes and workshops using design tools and design methods with people.

One of the first tools was the event "C'è Spazio per Tutti" (There's room for all) in November 2011 and in October 2012. It involved more than 50 international students from the M.Sc. in Product Service System Design and was visited by almost 550 citizens. Students developed a set of toolkits through PAR (Participatory Action Research) with the aim of enabling local residents to use the campus (a public space) by testing the kits directly with the users on the same day. "C'è spazio per tutti" was a way to test the developed research activities and to refine them to create an action format to be used later called the 'Social Innovation Journey (SIJ)' (Fassi et al. 2013). It systematizes activities recurring in the research projects and tries to help designers understand the stage of the social innovations they are dealing with and the potential of subsequent ones. The method leads to the stage of incubation of a social enterprise.

The wider role of design as a tool for developing policy agendas was proposed and carried out within CCAlps—the Creative Companies in Alpine Space project, which started in October 2011 and was completed in December 2014. The project engaged a network of nine partners from six different European countries and was mainly focused on regional planning for several reasons (Sedini 2015). First, the municipal level is less representative of the competition in advanced economies than regional authorities (Bontje and Kepsu 2013). Second, municipal borders do not define the limits of development of creative industries and activities, which is why gentrification of economic activities has been discussed (Smith and Williams 2013). Third, cities are becoming less appealing, particularly to creative workers within their peer group and or with their own family (Anzoise and Sedini 2011).

The different stages of the research focused on the development of entrepreneurial, creative ideas and, as a result, developed a set of policy recommendations to support the Cultural and Creative Industries. In particular, we identified a list of ten suggestions that were delivered to the Lombardy Regional Administration, which were shared with the general public during a final international event called "Cross Creativity". Among these recommendations, apart from those practically oriented to support CCIs (such as access to credit or skills development), there were also suggestions focused directly on the policy. We proposed, for example, the harmonization of the legal and legislative frameworks about CCIs, as well as the integration of policies and tools of intervention, internal to the various political and administrative sectors (e.g. the Departments of Culture, Research and Production Activities). During the project, a model of an intensive workshop for idea development called "Creative Camp" was tested and verified. We then held ten local and one international "Creative Camps", which produced almost 100 creative ideas. In order to encourage networking activities and to disseminate our results, we organized several public events, which attracted and involved approximately 4000 participants, including the general public and stakeholders. Stressing the importance of institutional involvement and its collaboration with enterprises, CCAlps amassed the participation of 240 institutions and 600 cultural and creative industries.

7.3 Resilience and Modernity in Milan

The concept of resilience seems to relate strongly to Modernity—or, more specifically, with the end of the modern era. Modernity was characterized by *grand* collective narratives (Lyotard 1984), which were also manifested through architecture and urban planning. During so-called post-modernism, individualism, liquidity and uncertainty took the place of these narratives (Bauman 2000). In this "liquid" society, life was precarious at the professional, political and personal level. Modernity and Post-modernity left very important legacies after they ended. Milan and other regions rearranged themselves with various contrasting results. This has been talked about as New Realism (Ferraris 2012, 2014) and Hypermodernism (Armitage 2000), when the most dangerous and uncertain elements of Modernity and Post-modernity are highlighted, such as the crisis of identity, the precariousness of life, closure and fear for personal security. In the conclusion of his book, *Milan since the miracle* (Foot 2001), the geographer John Foot stated that that Milan had lost its culture and its civic-mindedness. However, where the former contrast between society and community seemed clear (Tönnies 1887), there had been a reaction, mainly from below and partially influenced by new technologies. The Network Society theorized by Manuel Castells (1996), greatly worried the tech-sceptics who hypothesized that proximity, face-to-face relationships and even cities were in danger because of it. Instead, the so called Network Society favoured the circulation of knowledge and information; traces of communities (Bagnasco 1999) were created and re-created, within the interstices of nets and of cities sharing the same values, knowledge and goals. At this time, there was renewed attention towards narratives, but at a smaller and more local scale; peculiarities and local authenticities re-emerged. These dynamics were problematic if they were not driven by openness, conviviality and sustainability.

Thanks to the activation and implementation of projects that expect the participation and involvement of citizens, design can assist in making this shift.

Engagement is becoming the keyword and a necessary practice for the achievement of high standards in quality of life (Fassi and Sedini 2017). Social capital, which can be considered one of the most valuable capitals for regions and communities (Bourdieu 1980; Granovetter 1983; Putnam 1993) can be build or re-build thanks to the resilience capacities of design practices. The construction and establishment of a renewed social capital in Milan should happen in public spaces, also through collective moments, supported by both physical and virtual networks. In recent years this is what has been happening. From a historical and political point of view this can be seen has a reaction to the liberal policies that characterized the previous 15 years of the city—a reaction that, as stated by d'Ovidio,[3] is not framed outside the market and promotes social change through inclusion and relationships. Several activities set up in the area of Milan can be referred to here, such as the work done by Agenzia X in the book *Re/search Milano* (2015), which reveals initiatives shifting from the underground section of the city (which has actually always been there) to the mainstream section of the city. We can also highlight the recovery of the farmsteads (*cascine*), which were carried out in order to give back these beautiful spaces to the local people, as a good example. These interventions also include the redefinition of the functions of the farmsteads, even through co-design and participatory design.

The social housing projects are other good examples of resilience and participatory practices spreading in the region; the objectives of these experiments are not only to provide access to otherwise inaccessible[4] real estate market, but also to regenerate more peripheral and disadvantaged areas of the city, which suffer from the very centralized identification of Milan only with its city centre.

Kitchen gardens are also becoming an interesting case to be studied as they are increasingly being used as educational tools in schools, and as instruments of empowerment, collaboration and improvement of urban areas. In this chapter we are going to look at the example of *Coltivando*, a project of and at the Politecnico di Milano.

Another final example of places to generate and nurture social capital in Milan are incubators, creative labs and co-working spaces. In particular, we would like to mention Fabriq, which was the first municipal incubator of social innovation and was intentionally located in Quarto Oggiaro, a deprived area of the city. Beyond incubating start-ups, Fabriq organizes training courses for local entrepreneurship and projects such as *Milano Young Citizens.*[5] Within this topic of entrepreneurship and creativity for regional enhancement we will also refer to the European project CCAlps, which had as lead partners Regione Lombardia and the Politecnico di Milano.

In conclusion, the metaphor used by Geoff Mulgan (Murray et al. 2010) to describe how Social Innovation works seems very useful. In it, he affirms that cities are like a beehive. The bees in the hive represent citizens who have ideas or potentialities that have not been expressed. In order to give them the opportunity to put these potentialities into practice, the bees need trees, which (in the metaphor) represent public and private institutions that own power and wealth.

---

[3]http://www.ilfattoquotidiano.it/2015/12/21/milano-rinasce-prima-di-expo-e-oltre-la-darsena-cosi-torna-a-competere-con-le-grandi-capitali/2190414/

[4]In 2015 the average price of a house in Milan was 3.426 euro/m^2. This price has decreased by 19.8% since 2007 (Idealista.it data). The website http://www.borsinoimmobiliare.it/indicates 217.845 euro as the average price for a house in Milan.

[5]http://www.fabriq.eu/myc/

In order to make Milan still more attractive, talented and welcoming, this connection between the trees and the bees has to be incentivized, encouraged and empowered and is explemplified in the following case studies:

7.4 Case Study 1: Coltivando, the Convivial Garden at the Politecnico di Milano

Coltivando is a community garden within the Bovisa Campus of the Politecnico di Milano. It has been developed as a design experiment supported by the School of Design, The POLIMI DESIS Lab (belonging to the international DESIS Network), the "Human Cities—Reclaiming the public spaces" project under the "Creative Europe" program and "Feeding Milan, energies for change" research project funded by Fondazione Cariplo, a Milanese foundation, which aims to improve the efficiency of the food chain in the Milanese region.

As a single project, Coltivando brings together the concerns of both of these programs. It is situated in the public university space of the Politecnico di Milano's Bovisa campus, which allows the local community to discover a public place previously hidden to them and helps people in the community to grow their own food.

Service design thinking was combined with a spatial design approach and was used to develop Coltivando. A community-centred design approach engaged various stakeholders within the university community as well as those of the local Bovisa neighbourhood in Milan. The Coltivando research group is working alongside a diverse array of stakeholders in the university, business, government and civil society. The garden project has been co-designed considering topics such as service model, governance model, education and programming model, and spatial design. The service model of the garden is based on a collaborative model of sharing responsibilities amongst the group. A project like Coltivando coupled with service design models helps to address the gap between knowing the problem of unsustainability and finding solutions for individuals, sustainable design practitioners, communities and government through sustainable everyday design thinking and implementation. This is an experiment of collaboration between service and spatial design to merge diverse members of the community, who live in the same place, by engaging them in designing solutions for resilience for a place that suffered through changes in use classification. In the second half of the twentieth century, the Bovisa District has been subject to great change due to the removal of almost all the industries where most of the citizens living in the neighbourhood worked. In the last 20 years, an improvement of public transportation and the building of new residential areas have brought new life to the neighbourhood, but public spaces like green areas or landscaped squares are still inadequate. The Milano Bovisa Durando campus of the Politecnico di Milano, hosting the School of Design, was built at the end of the 1990s on the grounds of "Ceretti & Tanfani", an established company that produced cable railways and made Bovisa a working class district. The place is an important part of the historical memory of the residents. Today it is a university campus where students can attend classes, go to the library, meet at the canteen and use the workshops. In spring and summer, many students sit outside and enjoy the sun while doing outdoor activities. No one apart from the university community uses this area, and it could be considered as a 'hidden' public space (Fassi et al. 2016). The campus ultimately remains as an "island for students" and most of the people who once knew it as a former factory do not even have the chance to see how it has transformed—not because they are not allowed to enter, but because they think it is for students and university staff only. The two types of 'users' (university community and citizens) have very few contact points in common and the Coltivando project is attempting to change this situation.

In November 2011, there was an event called "C'è spazio per tutti" where professors, researchers and more than 60 students tried to figure out which design solutions could open up the campus to the neighbourhood. One of them was a garden bed where students engaged visitors to let them know how to interact with it, and which development potentialities could be done. Following this, a design research team was established. Sanders and Stappers (2008, 2012) state that 'co-creation practiced at the early front end of the design development process can have an impact with positive, long-range consequences.' According to that and after few months, three co-design sessions were run: two of them were addressed to the local inhabitants, and one of them was more for those people who were working in the university. The community-centered design approach was used to engage various stakeholders in the university community as well as in the community of Bovisa, and several tools were designed to let the people interact and design the garden.

The workshops included activities about roles and rules for the community garden and the development of spatial layout proposals. The researchers of the POLIMI DESIS Lab developed some service and spatial design tools based on role-play, "what if" techniques, surveys, ice breaking activities and basic elements for sketching. Attendants were split in groups by asking them to classify themselves as 'experts' or 'beginners' and asking them to tell their needs, expectations and skills. Workshops were aimed at defining a kind of zoning of the garden, which included tools space, herbs area, relax locations, etc., and were used to better understand the tasks for members and the rules to be followed. After the workshops, the researchers used the results to draw a very first design proposal both for the service model and for the space. When the proposal was verified with university logistic management of the campus, it was presented to the university administration to be funded. Coltivando took over 12 months to develop by a group of people who, every Saturday, spent their time building the DIY garden beds by assembling prefab steel panels, digging channels for the 450 m of tubes for the irrigation system, and putting 90 tons of organic soil in the garden beds.

Today, Coltivando is a community garden made of 100 garden beds containing more than 50 different vegetables and fruits and is managed by a team of 15–20 people both from the neighbourhood and the university. This team met regularly on Saturdays to work and spend time together. The garden is now also recognized as a place in the neighbourhood where people meet and organize events. This is slowly changing the perception of this public space—it is now less hidden and more open to all.

7.5 Case Study 2: CCAlps, Creative Companies in Alpine Space

CCAlps—Creative Companies in Alpine Space was a 3-year project financed within the Alpine Space Program of the European Union, concluded in December 2014. The focus was on the so-called Creative and Cultural Economy with the aim of improving the competitiveness and attractiveness of the Alpine Space Area for these sectors. The research phases led to the development of a model, composed of a set of objectives and tools, which produced some unexpected positive results.

The partners involved were both universities and institutions (e.g. Chambers of Commerce, regions) and also included nine Development Agencies from six different European countries (Italy, France, Germany, Austria, Slovenia and Switzerland). This composition of the research team was due to the specific focus on practical activities oriented to their translation into policy actions (Fassi and Sedini 2017).

CCAlps was based predominantly on the collaboration between institutional and governmental subjects, academia and creative and cultural enterprises. In particular, CCAlps' focus was oriented to the selection, development and realization of economic creative ideas with three main goals where design methods were fundamental, which were as follows (Sedini 2015):

1. *Support*: promote creativity and innovation as strategic factors for the development of the production activities; select and mentor ideas of business in cultural and creative sectors.
2. *Network*: facilitate the meeting between CCIs and other micro, small and medium enterprises (MSME), universities and research centres and stakeholders. In addition to that, create a transnational network among HUBs (incubators, service centres and co-working centres) that are present in the partners' regions, enterprises and among other stakeholders.
3. *Awareness*: increase the understanding and recognition of public institutions, stakeholders and SMEs about the role of CCIs.

The pilot action, which characterized supporting activities, was principally the Creative Camp. Through two versions of the Creative Camp, we were able to test and verify procedures that were focused on the support of early stage start-ups in cultural and creative sectors. The Creative Camp methodology wanted to capitalize on some similar activities done by the CCAlps' partners (design workshop, design jam sessions, call for ideas etc.) and then to take it a step further to implement a set of different, combined activities, such as: scouting of innovation ideas and projects, development of international mobility, mentoring to support the development of start-ups and networking among public events (Vignati 2015).

A Creative Camp can be seen as an advanced workshop, which has a very intensive initial phase of concept generation followed by a second, longer phase of idea development. Creative Camps included many activities to develop new products and services, improving the local productive system. During a period of 6 months in 2013 all the partners had to organize and hold their own Creative Camp. A general structure was provided to the partners, who could then organize and manage their camps on issues considered more "suitable" and appropriate to their regions and competences.

While these Creative Camps were structured in different ways according to the different regions, all of them had to follow some specific steps.

The first step was the launch of a call for ideas, which for the Lombardy region (Italy) was divided along three lines: multimedia and communication, fashion and service design.

The second step addressed the selection of the best ideas and 2 days of intensive workshop for concept development. Each region identified criteria on which basis the ideas were selected. Lombardy selected 25 ideas on the basis of the following criteria:

- *Usefulness*: the idea has to answer a question, a demand, an evident or hidden need of society (Woodman et al. 1993). In the CCAlps project, one of the criteria was the usefulness of the project, especially from a social point of view. For this reason, some projects displayed a more local angle because they referred to a specific context instead of other projects that could work across diverse social contexts;
- *Innovativeness*: having the OECD definition of innovation as a point of reference, according to which innovation is an implementation of new or better: the ideas to be selected had to constitute an element of change of already used practices in order to answer to the same need within the sphere of products and services, processes, marketing strategies, managerial strategies, workplaces, and or relationships,
- *Replicability*: the idea has to be reproducible. The criterion of replicability is closely related to the possibility of success. In order to achieve this standard it is necessary that the objectives and the expected results are declared, understandable and timely. Moreover, the process, in the first place, must possess the characteristics of adaptability and elasticity so as to be "exported" into other contexts.
- *Sustainability*: the idea must be sustainable (Diehl and Stroebe 1987). The sustainability of the idea can (and should) be expressed from an economic, an environmental and a social point of view.

The selected participants were then involved in a review phase. They had to actively work on their ideas, *re-generating* them according to the suggestions of experts and mentors. Then, going through a concept development phase, the participants re-framed their initial ideas and presented them to the experts involved. The participants had to explain their ideas again in a final presentation, during which the experts evaluated them again and selected the ideas to be admitted to the next step.

The third step was about mentoring and coaching. Mentors were identified according to the contents and the needs of the ideas selected. On-going reviews were implemented for some specific workshops with external experts about web design and business plan development. Three different tools were provided and set up (Vignati 2015):

- Qualification: specific learning activities were planned, starting from the analysis of the needs and the skill gaps of each team selected;
- Business Model Generation: this tool was proposed to all the teams;
- Cross Disciplinary Mentoring: the use of complementary skills (design, marketing, business) was crucial.

The fourth step was the organization of an international event. In the Lombardy event the mentored ideas had to make a creative pitch and a selected jury chose further ideas (six in total) which then had the opportunity to continue to be assisted in their development.

The fifth step was again about mentoring, finding possible partners and developing the most practical aspects of the idea. During this phase the international Creative Camp was held. The Creative Camp 2014 was considered as a further development of the projects in light of the final event in Milan. After the second Creative Camp, which was held on Lake Constance, 17 projects were selected to participate in the final event.

The sixth and final step was the organization of a common final event held at Regione Lombardia. The event, named Cross Creativity, was dedicated to cultural and creative start-ups and brought together over 300 start-ups. It was an international opportunity, realized under the patronage of Regione Lombardia and CCAlps project. Five out of the original 25 ideas from Lombardy, reached this final step and three of them have been created as real projects and are either already on the market or are about to be, which include:

1. Case Sparse|Tra l'Etere e la Terra (Spread Houses|Between Ether and Earth): a 3-year project which wants to discover and value remarkable areas thanks to the use of contemporary art. Link: http://www.casesparse.org/
2. MakersHub Milano: is a co-making and co-working space for makers, designers, DIY lovers and enterprises. It is a place for developing innovative products based on the interaction between craft and new technologies. Link: http://www.makershub.it/
3. Craftventure: a service which allows young people and tourists to experience artisans' work. At the same time, the artisans can preserve/renovate/transfer their knowledge thanks to the cultural "clash". This project won the contest during Cross Creativity. Link: http://www.craftventure.com/en/

All the activities carried on during CCAlps in general and the Creative Camp in particular were subjected to evaluation and analysis (carried out by Eupolis). A document of Policy Recommendation for the enhancement and the institutional support to CCIs was delivered to all the regions and representative institutions involved.

7.6 Case Study 3: ARNOLD

"ARNOLD—Art and Design in NoLo Social District" was a research and didactic activity with an in the field output done in 2016–17 by the POLIMI DESIS Lab, the NoLo Social District (Milan) and 50 students belonging to the Final Design Studio at the MSc Interior Design at the Politecnico di Milano. The main goal was to highlight some emerging features of the NoLo (North of Loreto) neighbourhood by using design methods. In particular, thanks to the collaboration of more than 4000 members of the local "social district", the goal was to create a diffused event where local artists (working or living in the neighbourhood) could exhibit their work in some unconventional places for art.

NoLo neighbourhood is a fertile ground for collaborative actions. Established in February 2016, the NoLo Social District is an online platform based on Facebook where neighbours could meet and ask for mutual aid on diverse issues. NoLo is a neighbourhood in Milan where a melting pot of cultures coming from several waves of immigration after the second world war led to an interesting platform for social cohesion and innovation activities. The online community organizes many offline activities that enable people to get in touch outside the internet (Pasqualini 2016) both to spend time together for the sake of it and to organize activities to improve the quality of the everyday life in the neighbourhood.

These offline gatherings usually start online by posting some ideas on specific interests and asking for other members interested in that to join. Many local initiatives have been established in this way—from the convivial breakfast on Saturday mornings, to the knitting club on Thursday evenings, from the establishment of a local radio network ("Radio NoLo"), to the happy hours with conversations in English and Spanish. Offline meetings are taking advantage of the neighbourhood public outdoor spaces or, when the weather gets cold, of the local pubs and bars who temporarily transform their spaces as locations for citizens meeting; Arnold was created following these untold rules. The Nolo Social district was an entrance gate for the research team to get in touch with the local artist and shop owners. Those places were belonging to the cultural network of the neighbourhood and chosen with the inhabitants: a parking area, a bar, a butchery, the local cinema, a printing shop, etc. (Fassi 2017).

Arnold was developed in two main phases: from July 2016 to February 2017 and from March 2017 to July 2017. In the first phase, the researchers team aimed to define design proposals through a co-design approach with artist, local shops owners, citizens and a group of students of the MSc Interior design of the Politecnico di Milano. The results were exhibited in February 2017 in the local indoor market where citizens were able to give feedback to be taken into account for the "action prototyping" of some solutions to project development and the final "mise-en-scene" schedule for June 2017 during a 3 days event. These phases belong to the PAR (Participatory Action research) methodology and creative placemaking approach to generate social innovation solutions.

The interaction among designers, artists, and locations owners created evidence of the different approaches to the creation of the output. If the artists roles were to give information to the designers to understand their "world", their artworks and how to give the right value to them in the exhibitions. Designers approach was led by the use of design tools and methods with a system approach (i.e. guided brainstorming, spatial journey maps, moodboards, technical drawings). At the same time the locations owners helped the designers and artists to make the exhibitions achievable (by offering their time, supporting them with small financial aid and let them temporarily change the layout of the shops) even if in some cases they put some constraints in the use of the spaces due to the small dimensions of them or to other ongoing activities.

Nevertheless, the actors involved went through a strong and inclusive collaboration for a transformative process and for (systemic) change in the neighbourhood. The combination of both research and practical actions could have potentially led to conflict since the roles of the members of the research team were different (Avison et al. 2007). This was taken under control due to a collaborative mood the actors held. The input designers and artists gave to the action prototyping was the shift from being object maker to maker of experiences (Spayde 2012) or better as 'city-makers'.

This city making was more evident in the second phase of the project, when the co-designed actions and exhibition went through the building phase. The assessment of the output of the first phase pointed out several interesting feedbacks. At first, the research team acknowledged that some actors collaborations were not efficient due to the different level of personal engagement. Some artists underestimated the time issue, for example, and were not able to dedicate an adequate time to the research teams. At the same time, after they realized how the exhibitions would have taken place and consequent temporary transformation of their locations, some location owners did not agree to participate anymore. It even happened that some artists were not happy about the output of the co-design activities and the exhibition settings, even if they were enthusiastic about the collaboration with the students including their approach.

Small-scale events in the neighbourhood originated within the community in response to a need or desire to celebrate their unique identity. The focus on the art through temporary and small-scale event in diffused location was a kind of answer to the need and desire by the neighbourhood itself to celebrate their unique identity (Douglas et al. 2001, p. 357). These kinds of events could even be seen as themed public opportunities to celebrate valued aspects of a community's way of life (Douglas et al. 2001, p. 358; Schlenker et al. 2005). In this case, the current low awareness in the neighbourhood of the cultural richness related to the ateliers and artists located in NoLo let the valued aspect be an emerging one.[6]

---

[6]The lack of awareness about the presence of artists, ateliers, and handicrafts in the neighbourhood came from some primary data collected through more than 50 surveys, video-interviews, informal chats done by the research team in 4 months time from September to December 2016.

"Social interactions between event visitors are an important part of the event experience and the level of satisfaction for the individual attending an event" (Nordvall et al. 2014). "Arnold" was designed not only to display pieces of art in unconventional places but also to improve the relationships among different and potential users (city dwellers, local inhabitants, occasional visitors, art-events fans). Since each location hosted a small activity to be carried out during the visit of the exhibition, Arnold had the double aim of connecting people to the artist's realm and having a personal experience through the visit.

The event was designed to have 13 locations with artists' exhibitions and two info-points,[7] it lasted for 3 days in June 2017 and had more than 1000 visitors. The success of the first edition put the premises to replicate the initiative by letting emerge how a temporary urban solution could take to a longer and more established one.

7.7 Conclusion[8]

In this section, we will summarize our observations and define the aims, drivers and tools that design is able to reach and activate to achieve regional enhancement.

According to the research activities shown above, we can state that designing for resilience aims at developing solutions for the following:

1. *Engagement of people from civil society and institutions*. (This first aim is directly connected with the achievement of those that follow.)
2. *Long-term economic strategies*. Even if the previously mentioned design practices were punctual and focused on the realization of specific events, it might be said that the sum of those repeated and cyclical events are able to set in place practices, which might not have an immediate result in terms of economic improvements but are able to activate other kinds of economies that are hidden. We can talk about the creation of a certain kind of environment (Marshall 1890; Becattini 1979; Santagata and Bertacchini 2011), also stated by theories such as Field-configuring events (Lampel and Meyer 2008; Sedini 2011); and that is able to influence and extend the consolidation of economies even not directly

[7]The involved artists and locations were: Mercato Comunale, Ghe Pensi M.I., Autorimessa Spoleto, with Barbara Colombo Vicolo del Fontanile (via Zuretti), with Andrea Salpetre @ZuArtDay & Looper Fest, Drogheria Creativa with Roberto Amoroso, Ci vuole un drink with Antonio Radice, Cinema Beltrade—with Patrizia Emma Scialpi, NoLo75 with Walter Paganuzzi (Timmerman Collective), Rovereto House & Lab with Lorenzo Picarazzi (Timmerman Collective),T12lab with Andrea Q, Officina del colore with Andrea Tarella, Carrozzeria 900, with Massimo de Caria, Angela Maria Capozzi, Grafica & Stampa Snc, with Eugenio Marongiu

[8]This chapter, and in particular the conclusions of this chapter, was preliminary to the writing of Fassi and Sedini paper "Design actions with resilient local communities: Goals, drivers and tools" published in 2017 on *Strategic Design Research Journal*, 10(1): 36–46, January–April 2017.

connected with business but which have economic fall backs in terms of facilitation and empowerment.

3. *Policy agendas*. The case studies examined in this chapter articulate two main ways in which design can contribute to the development of specific policy agendas. A first, direct and explicit contribution can be obtained through the cooperation of institutions in research projects aimed at developing and delivering policy recommendation (as with the CCAlps projects). The second contribution is indirect and it is carried out thanks to the capacity of certain research projects point out new and specific scenarios and views. In this last case, the usefulness of the research results depends on the attention and the willingness of institutions to listen and to take into considerations these insights for their policy agendas (as in the case of the Coltivando project) (Fassi and Sedini 2017).

Looking at the projects that were clearly starting from a specific area of interest and those that were open to the collection of new ideas from the population, we could identify specific drivers, which, in connection with design, seem to be at the cutting edge and were useful for regional resilience practices. These drivers were as follows:

- *Craft*: we noticed a renewed interest in traditional types of work, such as craft. This area of interest received attention not only in the urban environment but also in the countryside and rural areas where, at least in Italy, those kinds of activities are still ongoing. The practice of craftsmanship can be able to create a positive cultural environment and to increase social networks (Sennett 2008).
- *Do-it-yourself*: partially in connection with the previous point, makers seem to be the protagonists of a new "revolution". This fashion is also influencing the composition and perspective of cities and places in general. The diffusion of hubs and co-working spaces has been giving the possibility to gain new knowledge and also to enlarge one's own social network and social cohesion in general (von Streit and Lange 2013; d'Ovidio and Ranci 2014)
- *Relationships*: the so called Social Economy[9] has developed in order to answer social issues, in place of a declining welfare system. Start-ups which operate in the field of Social Economy are focus on sharing, collaboration, isolation reduction, taking care of common goods, etc. Calls for ideas, such as Chefare[10] in Italy, are specifically oriented towards the development of Social Innovation projects.
- *Arts*: in our examples of research there weren't many projects that focused on artistic matters. However, we think that this is an interesting emerging focus. Indeed the arts can also have an instrumental role for the promotion of the regions, the interpretation of regional value, the involvement of local communities, and the development of sustainable forms of tourism, etc. Also of real interest is the chance of cross-fertilization with other industries, such as ICT, which the arts sector allows (Throsby 2008).

[9]Here the definition given by OECD http://www.oecd.org/cfe/leed/social-economy.htm

[10]https://www.che-fare.com/

- *Cultural Heritage*: another interesting issue, which emerged in the projects described above, is the role of Cultural and Creative Industries (CCIs) in supporting cultural heritage. Indeed, CCIs can play a very important role in tourism and regional marketing. In particular, the importance of collaboration between museums and creative industries or the rise of so-called Creative Tourism defined as a tourism, which "*offers visitors the opportunity to develop their creative potential through active participation in learning experiences which are characteristic of the holiday destination where they are undertaken*" (Richards and Raymond 2000).

To deal with these drivers and to reach these goals, design could use several tools as explained in the previous case studies, which are summarized as follows:

- *Co-design workshops*: inhabitants are considered privileged witnesses, carrying out the real knowledge about a place or a situation. For this reason, engaging them in co-design workshops in very important to get useful information (and envisioning possible solutions), from one side, and to empower inhabitants thanks to design strategies, from the other side (Visser et al. 2005; Sanders and Stappers 2012);
- *Prototyping events*: the so-called Participatory Action Research (PAR) is constituted by a one-day event where ideas are tested by people (as users) using design toolkits. This kind of rapid small design experiments allowed immediate conclusions to be reached and then continued towards more stable and organized solutions (Fassi et al. 2013);
- *Calls for projects*: the collection of several ideas (and different point of views) on a same topic allows the creation of a panel of proposals. Going for quantity, in a first stage, allow to compare ideas according to specific indicators, such as innovation and scalability, and then select the best ones;
- *Social media strategy*: the dissemination of results is a very important step; indeed, from one side it allow to scale up the use of a specific solution; from the other side, it helps to 1) raise awareness towards specific issues; 2) spread information on design actions in the local context; 3) reach high numbers of people.

References

AaVv. (2015). *Re/search milano. Mappa di una città a pezzi*. Milano: Agenzia X.

Armitage, J. (Ed.). (2000). *Paul Virilio: From modernism to hypermodernism and beyond*. London: Sage.

Avison, D., Baskerville, R., & Myers, M. D. (2007). The structure of power in action research projects. In N. Kock (Ed.), *Information systems action research*. Boston: Springer.

Bagnasco, A. (1999). *Tracce di comunità*. Bologna: Il Mulino.

Bauman, Z. (2000). *Liquid modernity*. Cambridge: Polity press.

Becattini, G. (1979). Dal settore industriale al distretto industriale. Alcune considerazioni sull'unità d'indagine dell'economia industriale. *Rivista di economia e politica industriale, 1*, 35–48.

Björgvinsson, E., Ehn, P., & Hillgren, P. A. (2010, November). Participatory design and democratizing innovation. In *Proceedings of the 11th biennial participatory design conference, Sydney, Australia*. New York: ACM.

Bontje, M., & Kepsu, K. (2013). Creative knowledge strategies for Polycentric City regions. In S. Musterd & Z. Kovàcs (Eds.), *Place-making and policies for competitive cities*. Oxford: Wiley-Blackwell.

Bourdieu, P. (1980). Le capital social. *Actes de la recherche en sciences sociales, 31*(1), 2–3.

Brown, T. (2009). *Change by design*. New York: Harper Collins.

Cantor, J. (1995). *Experiential learning in higher education: Linking classroom and community*. Washington, DC: ERIC Digest.

Castells, M. (1996). *The rise of the network society: The information age: Economy, society and culture* (Vol. 1). Oxford: Blackwell.

Christopherson, S., Michie, J. E., & Tyler, P. (2010). Regional resilience: Theoretical and empirical perspectives. *Cambridge Journal of Regions, Economy and Society, 3*, 3–10.

Colucci, A. (2012). *Le città resilienti: approcci e strategie*. Pavia: Jean Monnet Centre.

d'Ovidio, M., & Ranci, C. (2014). Social cohesion and global competitiveness: Clustering cities. In C. Ranci, T. Brandsen, & S. Sabatinelli (Eds.), *Social vulnerability in European cities: The role of local welfare in times of crisis*. Basingstoke: Palgrave McMillan.

Diehl, M., & Stroebe, W. (1987). Productivity loss in brainstorming groups: Toward the solution of a riddle. *Journal of Personality and Social Psychology, 53*(3), 497.

Douglas, N., Douglas, N., & Derrett, R. (2001). *Special interest tourism: Context and cases*. Milton, Australia: Wiley.

Ehn, P. (2010). Participation in design things. In *10th biennial participatory design conference proceedings, Bloomington, Indiana, USA*. New York: ACM.

Etzkowitz, H., & Leydesdorff, L. (1997). *Universities and the global knowledge economy*. London: Pinter.

Etzkowitz, H., & Leydesdorff, L. (2000). The dynamics of innovation: From National Systems and "mode 2" to a triple Helix of university–industry–government relations. *Research Policy, 29*(2), 109–123.

Fassi, D. (2017). Events and the city: When Arnold meets NoLo. In D. Fassi & B. Camocini (Eds.), *In the neighbourhood*. Milano: Franco Angeli.

Fassi, D., & Sedini, C. (2017). Design actions with resilient local communities: Goals, drivers and tools. *Strategic Design Research Journal, 10*(1), 36–46.

Fassi, D., Meroni, A., & Simeone, G. (2013). Design for Social Innovation as a form of design activism: An action format. In *Social Frontiers: The next edge of social innovation research conference proceedings*. London: NESTA. Accessed December 10, 2017, from http://www.scribd.com/doc/191848489/Design-for-social-innovation-as-a-form-of-designing-activism-An-action-format

Fassi, D., Galluzzo, L., & Rogel, L. (2016). Hidden public spaces: When a university campus becomes a place for communities. In P. Lloyd & E. Bohemia (Eds.), *Proceedings of DRS2016: Design + research + society - future-focused thinking* (Vol. 8, pp. 3407–3421).

Ferraris, M. (2012). *Manifesto del nuovo realismo*. Bari: GLF editori Laterza.

Ferraris, M. (2014). Che cosa c'è di nuovo nel "nuovo realismo"? *Bollettino Filosofico, 27*, 119–126.

Fiksel, J. (2003). Designing Resilient, Sustainable Systems. *Environmental Science & Technology, 37*, 5330–5339.

Foot, J. (2001). *Milan since the miracle: City, culture and identity*. Oxford: Berg.

Frayling, C. J. (1993). Research in art and design. *Royal College of Art Research Papers, 1*(1), 1–5.

Granovetter, M. (1983). The strength of weak ties: A network theory revisited. *Sociological Theory, 1*(1), 201–233.

Graziano, P. (2012). Rischio, vulnerabilità e resilienza territoriale. *Quaderni del Dipartimento di Scienze Economiche e Sociali*, Serie rossa: Economia. Accessed December 10, 2017, from http://dipartimenti.unicatt.it/dises-wp_rossa_12_87.pdf

Hahlweg, D. (1997). The City as a family. In S. H. Crowhurst Lennard, S. von Ungern-Sternberg, & H. L. Lennard (Eds.), *Making cities livable. International making cities livable conferences*. California: Gondolier Press.
Hallegatte, S. (2014). *Economic resilience: Definition and measurement*. Policy research working paper, (No. 6852). Washington, DC: World Bank. Accessed December 10, 2017, from https://openknowledge.worldbank.org/handle/10986/18341
Lampel, J., & Meyer, A. D. (2008). Guest editors' introduction: Field-configuring events as structuring mechanisms: How conferences, ceremonies, and trade shows constitute new technologies, industries, and markets. *Journal of Management Studies, 45*(6), 1025–1035.
Lyotard, J. F. (1984). *The postmodern condition: A report on knowledge*. Minneapolis: University of Minnesota Press.
Meroni, A., & Sangiorgi, D. (2011). *Design for services*. Aldershot: Gower Publishing.
Manzini, E. (2015). *Design when everybody designs*. Boston: MIT Press.
Marshall, A. (1890/2013). *Principles of political economy*. New York: Palgrave Mcmillan.
Murray, R., Caulier-Grice, J., & Mulgan, G. (2010). *The open book of social innovation*. London: National Endowment for Science, Technology and the Art. Accessed December 10, 2017, from http://youngfoundation.org/wp-content/uploads/2012/10/The-Open-Book-of-Social-Innovationg.pdf
Nordvall, A., Pettersson, R., Svensson, B., & Brown, S. (2014). Designing events for social interaction. *Event Management, 18*(2), 127–140.
Pasqualini, C. (2016). Una nuova cultura della socialità: la sfida delle social street a Milano. In R. Lodigiani (Ed.), *Milano 2016. Idee, cultura, immaginazione e la Città metropolitana decolla*. Milano: Franco Angeli.
Pinto, H. (2015). *Resilient territories: Innovation and creativity for new modes of regional development*. Cambridge: Cambridge Scholars Publishing.
Pisano, U. (2004). Resilience and sustainable development: Theory of resilience, systems thinking. *European Sustainable Development Network (ESDN), 26*, 50.
Putnam, R. D. (1993). *La tradizione civica nelle regioni italiane*. Milano: Mondadori.
Richards, G. W., & Raymond, C. (2000). Creative tourism. *ATLAS news, 23*, 16–20.
Rodin, J. (2014). *The resilience dividend: Managing disruption, avoiding disaster, and growing stronger in an unpredictable world*. London: Profile Books.
Sabatino, M. (2015). *Competitività e Resilienza dei Distretti Produttivi in Sicilia*. XXXVI Conferenza Italiana di Scienze Regionali, Cosenza. Accessed December 10, 2017, from https://www.researchgate.net/profile/Michele_Sabatino/publication/281836669_Competitivita_e_resilienza_dei_Distretti_Produttivi_in_Sicilia/links/55fa899208ae07629e0215d7/Competitivita-e-resilienza-dei-Distretti-Produttivi-in-Sicilia.pdf
Salzano, E. (1997). Seven aims for the Livable City. In S. H. Crowhurst Lennard, S. von Ungern-Sternberg, & H. L. Lennard (Eds.), *Making cities livable. International making cities livable conferences*. California: Gondolier Press.
Sanders, E., & Stappers, P. (2008). Co-creation and the new landscapes of design. *CoDesign: International Journal of CoCreation in Design and Arts, 4*(1), 5–18.
Sanders, E., & Stappers, J. P. (2012). *Convivial toolbox. Generative research for the front end of design*. Amsterdam: BIS Publisher.
Santagata, W., & Bertacchini, E. (2011). *Creative atmosphere: Cultural industries and local development* (No. 4/2011). Working paper Department of Economics "S. Cognetti de Martiins". Accessed December 10, 2017, from https://www.css-ebla.it/wp-content/uploads/4_WP_Ebla_CSS-2.pdf
Schlenker, K., Edwards, D. C., & Sheridan, L. (2005). A flexible framework for evaluating the socio-cultural impacts of a small festival. *International Journal of Event Management Research, 1*(1), 66–77.
Sedini, C. (2011). Field-configuring events: Design fairs and cities. *REDIGE Edição Especial: Economia Criativa, 2*(2), 64–78.

Sedini, C. (2015). Don't reinvent the wheel: A design driven agenda for cultural and creative industries. In A. Meroni, L. Galluzzo, & L. Collina (Eds.), *Proceedings conference, the virtuous circle. Design culture and experimentation* (Politecnico di Milano, 3–7 giugno 2015). Milano: McGraw-Hill Education.

Sennett, R. (2008). *The craftsman*. London: Penguin.

Smith, N., & Williams, P. (2013). *Gentrification of the City*. London: Routledge.

Sotarauta, M. (2005). Tales of resilience from two Finnish cities, self-renewal capacity in the heart of strategic adaptation. In C. Duke, M. Osborne, & B. Wilson (Eds.), *Rebalancing the social and economic learning, partnership and place*. Leicester: Niace.

Spayde, J. (2012). Public art and place making. *Public Art Review, 47*, 22–25.

Stumpo, G., & Manchin, R. (2014). *The resilience of employment in the Culture and Creative Sectors (CCIs) during the crisis*. EENC Report. Accessed December 10, 2017, from http://www.interarts.net/descargas/interarts2575.pdf

Throsby, D. (2008). The concentric circles model of the cultural industries. *Cultural trends, 17*(3), 147–164.

Tönnies, F. (1887/2005). Community and society. In J. Lin & C. Mele (Eds.), *The urban sociology reader* (pp. 16–22). New York: Routledge.

Vignati, A. (2015). Entrepreneurship in the creative and cultural sector. In A. Meroni, L. Galluzzo, & L. Collina (Eds.), *Proceedings conference, the virtuous circle. Design culture and experimentation* (Politecnico di Milano, 3–7 giugno 2015). Milano: McGraw-Hill Education.

Visser, F. S., Stappers, P., Van der Lugt, R., & Sanders, E. B. (2005). Contextmapping: Experiences from practice. *CoDesign, 1*(2), 119–149.

Von Streit, A., & Lange, B. (2013). Governance of creative industries: The role of social and professional networks. In S. Musterd & Z. Kovàcs (Eds.), *Place-making and policies for competitive cities*. Oxford: Wiley-Blackwell.

Walker, B., Holling, C. S., Carpenter, S. R., & Kinzig, A. (2004). Resilience, adaptability and transformability in social–ecological systems. *Ecology and Society, 9*(2), 5.

Woodman, R. W., Sawyer, J. E., & Griffin, R. W. (1993). Toward a theory of organizational creativity. *The Academy of Management Review, 18*(2), 293–321.

Chapter 8
Organisational Innovations for Science-Industry Interactions: The Emergence of Collaborative Research Centres in Spanish Regional Innovation Systems

Sandro Giachi and Manuel Fernández-Esquinas

8.1 Introduction

The aim of this chapter is to analyse the creation and diffusion of organisational innovations for knowledge transfer in regional innovation systems. An important example of these innovations is found in the collaboration between scientific and industrial sectors. We study a specific case of organisational innovations: collaborative research centres (CRCs) constituted by the joint participation of scientific, government, and industrial agents. These organisations tend to emerge from programs undertaken frequently by regional governments or other local agents, with the goal of increasing the technical scientific capacities of the territory and guiding them towards strategic sectors for the development of the productive system. It can therefore be argued that the creation and diffusion of CRCs is a phenomenon relevant to understanding the innovative dynamics associated with regional resilience.

The establishment of new innovative agents through intersectoral collaboration between science, industry, and government could have an impact on those factors that characterise the resilience of a region: industrial composition, knowledge networks, and institutional structures (Boschma 2015; Pinto and Pereira 2013). The creation and diffusion of entities such as CRCs tend to involve a restructuring of networks of existing relationships in a region, as well as a gradual transformation of the strategies or the institutional structure of the organisations involved in the

S. Giachi (✉)
Science Policy Research Unit (SPRU), University of Sussex, Brighton, UK
e-mail: s.giachi@sussex.ac.uk

M. Fernández-Esquinas
Institute for Advanced Social Studies (IESA), Spanish National Research Council (CSIC), Cordoba, Spain
e-mail: mfernandez@iesa.csic.es

H. Pinto et al. (eds.), *Resilience and Regional Dynamics*, Advances in Spatial Science, https://doi.org/10.1007/978-3-319-95135-5_8

collaboration (e.g. universities, companies). In some cases, CRCs make possible the productive specialisation of a conglomerate of the region's firms, or help to identify new strategic areas for research and production activities.

Due to the lack of studies about regional resilience from an organisational innovation point of view, we use an original perspective to understand the transformation of the regional innovation systems. Typically, the studies that associate regional resilience with innovative dynamics focus on aspects such as creating intellectual property, adopting new technologies, or introducing new processes or products on the market. Through our research, we demonstrate the benefit of analysing a region's resilience through its capacity for restructuring linkages between agents and for establishing new proactive endeavours in knowledge production processes and technological innovation through specialised organisations such as CRCs.

We analyse the process of creating and diffusing CRCs in innovation systems using a case study of a country (Spain) and its regional systems. For this purpose, we describe the evolution of the Spanish R&D system and the impact of the recent economic crisis. We then discuss the existing gap between science and industry in the Spanish R&D system and the emerging initiatives for counteracting it through organisational innovations, mainly by the regional governments.

Our methodology is based on an exhaustive map of existing centres in Spain that was created using a heterogeneous set of documentary and secondary data sources (evaluation reports and results, statistics elaborated by official entities, science plans, technology, innovation and industry, R&D programs, corporate reports, institutional directors, and web pages). This information was completed with studies of the Spanish case available in the bibliography, as well as data from a survey addressed to research centres in 2012. These data sources allow us to describe the trends in the Spanish R&D system that are related to the organisational innovations for science-business relations.

The structure of the chapter is as follows. In Sect. 2, we give a brief introduction on the concepts used. In Sect. 3, we present an overview of the Spanish R&D system, focusing our attention on its recent evolution and those aspects of the system that make collaboration between science and industry difficult. In Sect. 4, we describe the public programs and regional initiatives for constructing organisational innovations geared towards linking science and industry. We continue with a description of the population from the map of CRCs and the results from a survey on the dynamics of collaboration and knowledge transfer between researchers and firms. In Sect. 5, we discuss the conclusions and implications for the dynamics of resilience and innovation in regional systems.

8.2 Organisational Innovation and Resilience in Regional Innovation Systems

Resilience is a concept (mostly imported from natural sciences) that refers to the capacity of a particular system to recover its equilibrium and long-term trajectory after experiencing a shock or crisis. This concept has been applied to regional studies from an evolutionary perspective (Pinto 2015). A resilient region is a territory with the potential or capacity for stable adaption when facing potential shocks to its technological, economic (market), or political structure that can negatively impact its trajectory as an evolving system (Simmie and Martin 2010; Martin and Sunley 2015). Therefore, a resilient region is able to adapt its economy in the long term and recover its growth trajectory, a competence that may include the introduction of new development paths and innovative solutions.

Recently regional innovation systems have been proposed as strategic places for the study of economic resilience and for identifying resilient regions (Pinto and Pereira 2013). It is argued that certain elements of the innovation system constitute an essential component of resilient economies (Christopherson et al. 2010; Simmie 2014). For example, the capacity of a territory that allow to recover its growth path and development after having experienced a shock or crisis in its technological, economic, or political trajectory can be found in in a specific type of innovation, some times of an organisational nature.

The concept of "organisational innovation" can be distinguished from other types of innovations (e.g. technological, social, or institutional) because it relates to the organisational sector (Lam 2005). This chapter uses the term "organisational innovation" to refer to the creation and diffusion of a new type of organisation or organisational model. It differs from other usual meanings (Hage 1999) related to "establishing a new way of working within an already existing organisation" or the capacity of an organisation to produce innovations. The concept of organisational innovation resembles "institutional innovation," although it is not exactly the same: when organisational innovations form a new population and are institutionalised, then they may be considered institutional innovations. They form new institutions understood as the evaluative and regulatory aspects that govern the life of the organisations and legitimise their operations. Similarly, in our perspective research we mention "innovation processes" and its two possible meanings: the creation of something new or its diffusion.

In the scope of innovation systems of peripheral regions, as is the case for some areas of Southern Europe, organisational innovations constitute a policy to facilitate science-firm relations (Fernández-Esquinas et al. 2012; Fernández-Zubieta et al. 2016). Commonly these innovation systems share rigid and bureaucratised organisational structures that become an obstacle for university-firm relations and knowledge transfer from the public science sector to the productive system and private industry. In these regional systems, scientific agents tend to focus their efforts on basic research activities guided by criteria of scientific quality and are financed mainly through public funds. An important part of the industrial tissue tends to work in traditional and low-intensity technology sectors, and seeks competitive advantages through the use of intensive labour and low production costs.

This social and economic structure of R&D systems of peripheral regions implies a barrier in knowledge transfer that has consequences for resilience during periods of economic crisis. The crises are especially hard on the system's public entities—universities and public academic centres: they suffer from a decrease in the availability of public funds, which cannot be easily substituted by R&D services from the industrial environment.

On the other hand, firms find it difficult to compete by using products with greater added value based on knowledge and a highly skilled workforce. The difficulty of accessing the sources of knowledge accumulated in the public sector is an important barrier. The periods of crisis can be seen as times at the structural limitations of a territory emerge. For R&D systems of peripheral regions, periods of crisis can imply a risk of widening the gap between public science and private industry.

In these cases, radical transformations in the organisation of R&D activities and interinstitutional relations are considered as a strategy to facilitate a greater linkage between the scientific and industrial sectors. CRCs are an example of organisational innovation capable of creating a bridge between different sectors and promoting knowledge transfer, technological innovation, and ultimately, socioeconomic development of a territory (Gray et al. 2013).

Recent studies (Boschma 2015; Pinto and Pereira 2013) have described some of the factors that characterise the resilience of a territory, such as industrial composition, knowledge networks, and institutional structures. Some of them are closely related to organisational innovations, like CRCs. The transformation of organisational relations and inter-organisational relations between science and firms have to do with the following factors. First, CRCs act upon knowledge networks because they promote both the consolidation of existing ties (regardless of the strength of the tie) and the creation of new linkages. In other words, they are mechanisms for creating social capital among agents of the regional innovation system (Garrett-Jones 2007). Second, CRCs can serve as a vector of transformation of existing institutional structures for various reasons. They guarantee more flexibility in the organisation of public research and they decentralise the resources and operations originating from the government sector, consistent with the governance paradigm of R&D and innovation (Clark 2010). Furthermore, they stimulate strategies of open innovation in private firms through collaborative knowledge management mechanisms (Perkmann and Walsh 2007). Third, CRCs can affect the productive structure and economic specialisation of a territory. For example, they can revitalise traditional sectors of the regional economy, open the way for the creation of new industrial clusters, or even identify strategic innovative areas for R&D in the territory (Liyanage 1995).

8.3 The Spanish R&D System During the Economic Turmoil

8.3.1 Size and Evolution of the System

Spain occupies an intermediate position among the global R&D systems. In 2014, internal R&D expenditures were 12.821 billion euros, which represents 1.23% of

GDP. This number puts Spain below the European average (2.03% of GDP for UE-28) and the OECD member countries (2.40%). It is located approximately in the third quartile of the rankings, along with Italy and Hungary (COTEC 2015; FECYT 2015). Investment in R&D activities in Spain is notably lower than it is in other European countries like Finland (3.6% of GDP), Sweden (3.4%), Germany (2.9%), and France (2.2%). Spain's innovative effort is modest even if we record it as *per capita* expenditures relative to the resident population.

Moreover, the figures in 2014 relative to total internal R&D expenditures reflects the downward trend that began in 2011 as a consequence of the current economic crisis (Fig. 8.1). Between 2008 and 2010 the absolute figures (more than 14.5 million euros) and relative figures (almost 1.40% of GDP) rose. However, since then they have decreased (INE 2015), halting the slow process of convergence with the European levels that had been generating since the end of the 1980s.

If we look at the composition of total internal R&D expenditures compared to GDP (Fig. 8.1), we see that the majority of investment comes from the private sector, which includes firms, private institutes and non-governmental organisations (NGOs), followed by the higher education sector (e.g. universities), and public administration. In 2014, the private sector represented 53.1% of total national internal R&D expenditures, compared to 28.1% from the higher education sector, and 18.7% from administration. However, the private sector's contribution to R&D expenditures in Spain is still lower than it is in other countries in our community (COTEC 2015; FECYT 2015).

Figure 8.1 also demonstrates the trend related to the increase in the weight of the private sector between 2005 and 2008, both in absolute and relative values. Subsequently, there was a minor but constant downward trend. In parallel, at the beginning

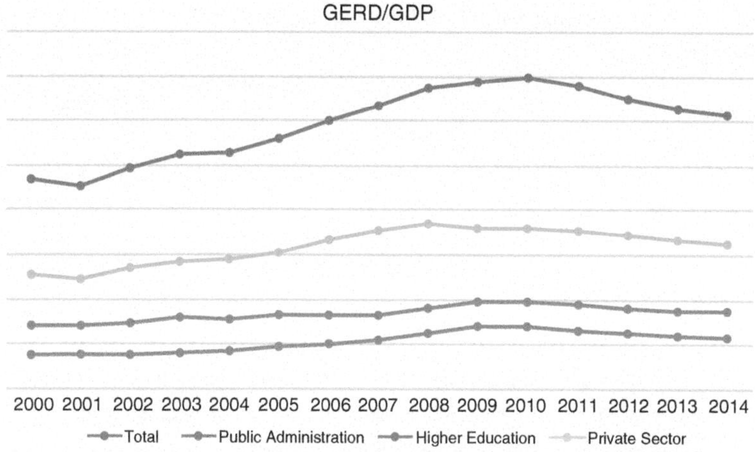

Fig. 8.1 Evolution of R&D expenditure in Spain by sector. Source: Spanish statistical office (INE 2015)

of this period (2008–2010) we observed a slight upward trend for the public sector (higher education and administration) that, however, was suddenly interrupted in 2011. Spain's current R&D investment levels have returned to approximately the same levels as in 2006.

With regard to the human resources employed, we see similar dynamics. The number of personnel employed in R&D activities in 2014 was 332,871 people, which in terms of full-time equivalents was 200,237 units, or 1.2% of the total employed population. Furthermore, 64% of personnel employed in the R&D sector were concentrated in universities and administration. Figure 8.2 shows the evolution of R&D personnel by sector. The number of R&D employees increased considerably until 2008, and then suffered a sort of stagnation and started to slowly decline in 2011. The private sector was the one that grew most since 2002, reaching its peak in 2008 and later decreasing, anticipating the subsequent trend in the public sector that, in turn, continued growing (although with less force) until 2010, and later decreased.

If we compare the data between regions in Spain, the regions that employs more personnel compared to the national total are Madrid (23.2%) and Catalonia (21.9%), followed by Andalusia (11.8%), Valencia (9.3%), and Basque Country (8.9%). This distribution has remained the same despite changes in the last several years, although the weight of Madrid and Catalonia is gradually decreasing (COTEC 2015). The number of researchers is proportional to the total R&D personnel between the regions (Fig. 8.3). The only exception to this trend is the slightly greater weight of the researchers compared to total R&D workforce found in Basque Country and the slightly lower weight researchers have, in comparison, in Andalusia.

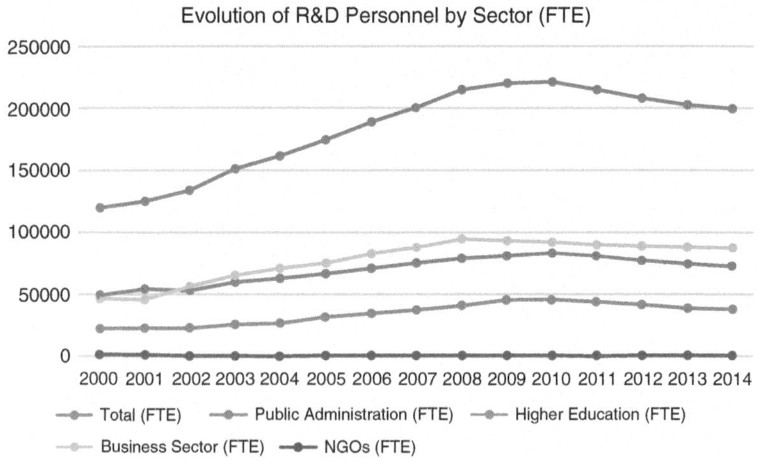

Fig. 8.2 Evolution of R&D personnel by sector. Source: Spanish statistical office (INE 2015). Numbers in Full-Time Equivalent (FTE)

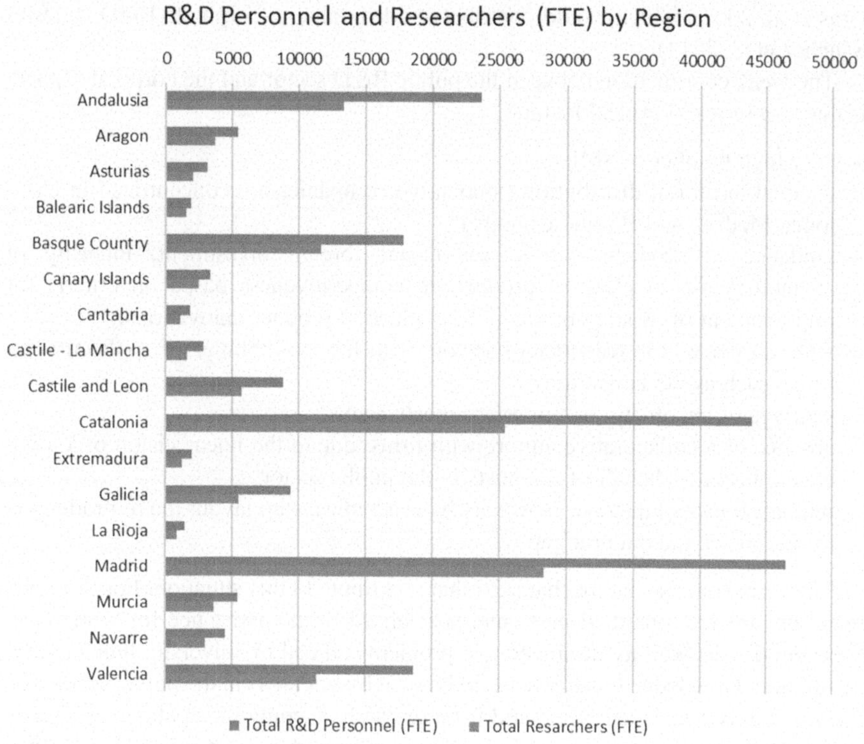

Fig. 8.3 R&D Personnel and researchers by region. Source: Spanish statistical office (INE 2015). Numbers in Full-Time Equivalent (FTE)

8.3.2 *The Science-Industry Gap in Spain*

According to the European Commission, Spain is a "moderately" innovative country (European Commission 2010a, b), characterised by weak science-industry relations, as well as relatively low levels of academic-based entrepreneurship and public-private collaboration in R&D.[1] In the Spanish case, it is argued that collaboration between science and industry would have beneficial effects on innovation and economic development. For example, incorporating scientific personnel from the public sector in firms could increase investment in internal R&D activities, and at medium-term, the propensity to register patents. Another useful issue is the creation of spin-offs, start-ups, and new businesses from the university environment (Gómez

[1]One example of this trend would be the number of public-private co-authorships in scientific publications. Despite having gone from 22.5 units (per one million individuals from the national population) in 2008 to 28.7 units in 2011, this figure is still very low compared to the UE-28 average, which is 52.8 units (European Commission 2010b).

Gras et al. 2007), which would contribute added value to corporate R&D (Iglesias Sánchez et al. 2014).

The weak coordination between the public R&D sector and the corporate sphere is due to a series of critical factors:

- very high number of SMEs;
- unequal territorial distribution (innovative companies are concentrated in Catalonia, Madrid, and Basque Country);
- guidance of productive processes (scant foreign investment, intensity of technology-use in place of production, competitiveness based on criteria for improvement of existing production, in place of product innovations);
- wide presence at universities of sectors with few possibilities of application by firms, such as the humanities;
- shortage of university-firm interface mechanisms;
- absence of a collaborative culture with firms, due to the linear vision of knowledge transfer on behalf of scientists in the public sector;
- mutual climate of distrust, especially by academics, who favour the re-production of this social and cultural gap.

There are some social mechanisms that contribute to this situation. For example, there are still few qualified personnel specialised in the assistance for working in these entities, as well as administrative problems related to university bureaucracy. In addition, knowledge transfer tends to be seen by scientists in the public sector as a process that is established afterward from research, keeping the contexts of discovery and application separate. In addition, an important part of university-firm interactions is due to the necessity of firms to solve concrete technical problems by using assistance or consultancy service.

The existence of this gap has not stopped firms in Spain from continuing to ask more intensely for highly qualified personnel for development of projects related to innovation or from believing that research contracted with the public sector is satisfactory (Mulet Meliá 2003). At the same time, collaboration with firms is believed to positively influence the scientific productivity of academic researchers (Manjarrés-Henríquez et al. 2009). On the other hand, universities are increasingly looking to approach the corporate sector, both to acquire financing (e.g. due to criteria established by competitive public calls) and to improve the recognition and social impact of their training and research programs, within the framework of the so-called "entrepreneurial university" paradigm and its growing legitimisation by national and supranational institutions (Díaz-Catalán et al. 2011; Palomares Montero et al. 2012).

Despite these limitations, there are two aspects that contribute to fuelling change in the existing model in the public and government sectors: financing science and managing R&D activities. Traditionally in Spain financing of R&D has been sensitive to economic circumstances. The public R&D budget tends to grow in the peak phases of the economic cycle but falls quickly when there is recession, losing part of the ground that had been gained before, due to relying heavily on public financing. This reflects the absence of a political consensus for maintaining R&D as a priority of the State (Vence and Heijs 2006). To this situation we must add the constant heterogeneity of guidance from the regional R&D policies (Sanz-Menéndez and Cruz-Castro 2005) that contribute to increasing the system's variability.

As a consequence of the continuous changes in the budgetary availability, along with the scarce margin of administrative manoeuvring due to the strict bureaucratic system, some governments and other public agents have started launching new types of actions, aimed at creating new organisational units capable of obtaining higher levels of external financing (e.g. private, international) and that would allow greater administrative flexibility in order to facilitate more efficient organisation of resources, the potential of concentrating efforts around a mission, a line of research, a technology, a specific project or a more effective and fluid management of cooperation between firms and other entities (Cruz-Castro and Sanz-Menéndez 2007; Arias Aparicio 2011; Fernández-Esquinas and Ramos-Vielba 2011).

Organisational innovations should be understood in the context of a multilevel government in which there are various competencies in the innovation policies, where actors from the general administration of the state, the regional governments, and the universities overlap. The most recent political initiatives come mainly from the regional spheres, as well as from the procedures of universities and public centres and some corporate associations. This involves procedures that try to get past the barriers for the transfer and collaboration stemming from the bureaucratic structure of the R&D system and from the traditional policies based on linear models of innovation.

8.4 Collaborative Research Centres in Spanish Regional Innovation Systems

8.4.1 The Role of Regional Innovation Policies in Organisational Innovations

Some regional governments have distinguished themselves by defining and executing programs and guided actions in order to favour collaborative research. In some cases, these initiatives are in agreement with national policies, while others adopt alternative or even conflicting approaches (Tortosa 2006). Overall, regional governments have been guided more towards the objectives of technology development and innovation support more than national government do. Examples are the creation of technology transfer offices, Science Parks, and technology centres (Sanz-Menéndez and Cruz-Castro 2005). The entirety of regional governments has R&D and innovation development plans, usually with significant support from European funds. In all of these plans, public-private collaboration is considered one of the desirable goals of the regional innovation system, although with significant variations (COTEC 2007).

In the review of the most recent innovation plans, there is a convergence on the objectives and tools proposed in order to promote collaborative research, but we also see different levels of intensity and involvement. For example, in some regions there are no obvious traces of specific actions in this sense, beyond the intent of creating these entities. In reality, R&D procedures focus on management of competence transfer in higher education and management of universities, with little budgetary margin for organisational innovations directed toward industry. In turn, other regional governments have been more active, like in the case of Asturias or Galicia, where a series of clearly defined entities has been created to strengthen science-industry collaboration, along with other actions aimed at creating technology clusters in regional strategy sectors (e.g. lumber, energy, environment, dairy) and strengthening those that already exist. We see something similar in Catalonia, in a more diverse manner due to the size of the system, through actions guided towards consolidating the regional CRC network, or towards promoting corporate innovation through cooperation between firms, technology centres, research centres, and universities. In other cases, the regional initiative has been found with the strategy of a large multinational corporation. This is the case of three innovation centres promoted by Microsoft in collaboration with the regional government of three Autonomous Communities: the Balearic Islands, Cantabria, and Catalonia. The goal of these initiatives is to make the most advanced Microsoft technological platform, solutions, and TIC tools available to firms and their employees in order to contribute to the growth of the local industry.

On the other hand, some regions have made specific tools and organisations available to promote stability of collaborative research. Andalusia presents a relevant case. In 2005 it created the *Corporación Tecnológica de Andalucía* [Technological Corporation of Andalusia] (CTA), a non-profit organisation whose objective is financing corporate R&D projects in collaboration with research groups from universities or from the public sector; half of the project financing is guaranteed by CTA, while the other half is provided by the firms (Fernández-Esquinas et al. 2012). A more extensive initiative, albeit less innovative from the point of view of cooperation with the university, was undertaken by the Community of Valencia, which, between 2001 and 2006, grouped the 14 existing Technology Institutes in the territory into one consortium in order to grant them with funds to favour providing advanced R&D services to SMEs in key industrial sectors within the region (e.g. ceramics, energy, food and agriculture, biomechanics, optics, footwear, textiles, toys, construction).

Finally, the Basque Country is the region where there has perhaps been the most diversification of initiatives and where programs have been launched that have a certain magnitude because its R&D policy has traditionally been linked more with business and oriented towards the creation of its own infrastructure of R&D agents (mainly technology centres). Once consolidated, that infrastructure was a starting point for promoting relations with the rest of the agents (firms, universities).[2] Among the different initiatives launched over the last several years by the Basque government, the following programs are of special interest for our research.

The Basque Excellence Research Centre (BERC) program, whose objective is to raise the level and quality of scientific research in the region, attracts "star" scientists from outside the Basque scientific system and creates new model research groups in its areas of knowledge. The four existing BERC centres support the region's university groups.

The Cooperative Research Centre (CRC) program's objective is to create centres that will act as innovation networks in equilibrium with the goal of scientific excellence and commercial marketing of the results. The knowledge generated by CRCs is aimed at particular spheres of research because of their contribution to key sectors in the economy (e.g. manufacturing technologies) or to stay in line with strategic diversification policies (e.g. biosciences and nanosciences) where the region has identified a significant potential for improvement or where it wishes to attract international research personnel. The six currently existing CRCs are autonomous entities, mainly concerned with personnel management and collaborative agreements.

Other regions have launched their own programs that anticipate ex novo creation of a network of institutes for collaborative research, financed both through regional and private funds. For example, in 2005 the Community of Madrid created the *Instituto Madrileño de Estudios Avanzados* [Madrid Institute of Advanced Studies] (IMDEA) in order to promote innovation activities in socially useful areas, to approach a critical mass of researchers and teams of international quality, and to attract firms and create a competitive environment based on the generation of knowledge. Its legal formula was that of a private non-profit organisation, which provided great flexibility and agility in management. Its internal operations in human resources, launching new initiatives, gaining external funding, opening new lines, and in research projects were similar to those of a private firm.

8.4.2 The Map of CRCs in Regional Innovation Systems

The process of mapping centres has allowed the identification of 216 existing collaborative research centres in Spain. Table 8.1 shows the distribution of the existing types of CRCs in Spain based on the official definition used to design them. We have identified at least two levels for classifying the centres according to the official denominations typically used, like in a taxonomy system. First we

[2]One example is the process of consolidation and agglomeration of ITCs in this region that ended in the creation of two large Technology Corporations: Tecnalia and IK4. Due to its size and critical mass, Tecnalia has become the first private centre for applied Research in Spain and the fifth in Europe, while IK4 is the European technology corporation with a higher ratio of revenue as a result of directly invoicing firms.

Table 8.1 Types of CRC

Official denomination of centre	N	%
Innovation and technology centres (ITCs)	139	64.4
FEDIT Centres (FEDIT: association of technology centres)	43	19.9
IK4 Centres	9	4.2
TECNALIA Centres	3	1.4
Microsoft innovation Centres	3	1.4
Other	81	37.5
Cooperative research and excellence networks	27	12.5
CIBER	9	4.2
IMDEA	7	3.2
CRC	7	3.2
BERC	4	1.9
Ad hoc R&D institutes	50	23.1
Semi-public Centres	23	10.6
University Institutes	11	5.1
IESE-based University Institutes	8	3.7
Semi-private Centres	8	3.7

Source: Map of centres (ES/CRC 2012); own elaboration

consider the most general level (in boldface, Table 8.1), formed by three categories: innovation and technology centres (ITCs), cooperative research and excellence networks, and *ad hoc* R&D institutes. Among these, ITCs constitute the largest category, occupying almost two-thirds of the distribution (64.4%), followed by *ad hoc* R&D institutes (23.1%), and centres created through programs for the development of collaborative research networks (12.5%).

If we look at the second level of classification, we see that the specific type that occurs most frequently in the distribution comes from entities registered as innovation and technology centres that are not part of the FEDIT or any other sectoral group: these 81 centres constitute 37.5% of the population (Table 8.1). On the other hand, the second largest category is formed by ITCs affiliated with FEDIT (43 cases, 19.9%), while the rest of the ITCs are composed of nine centres that form part of the Basque group IK4, the three centres from the Basque group TECNALIA, and the three Microsoft innovation centres working in collaboration with regional governments.

As we have seen, there are only four public programs in Spain that have led to creation of a network of collaborative research centres. Among these, the largest group (Table 8.1) is made up of the nine *Centros de Investigación Biomédica En Red* [Biomedical Research Networking Centres] (CIBER, per Spanish acronym); these are followed by the seven Madrid Institutes of Advanced Studies (IMDEA) and other CRCs from Basque Country. Finally, there are the four Basque Excellence Research Centres (BERC). Among *ad hoc* R&D institutes, the largest group is composed of semi-public centres (23 cases, 10.6%), followed by the *Institutos Universitarios de Investigación* [University Institutes for Research] (IUI, per

Table 8.2 Organisational age by type of CRC

V = 0.281 (P value <0.001)	Taxonomy ITC (%)	Networks (%)	Ad hoc (%)	Total (%)	Cumulative (%)
1–5 years	19.4	33.3	24.0	22.2	22.2
6–10 years	23.0	63.0	30.0	29.6	51.9
11–15 years	18.7	3.7	28.0	19.0	70.8
16–20 years	13.7		6.0	10.2	81.0
20+ years	25.2		12.0	19.0	100.0
Total	100.0	100.0	100.0	100.0	

Source: Map of centres (ES/CRC, 2012); own elaboration

Spanish acronym) with public-private participation (11 cases), the eight collaborative institutes supported by the *Instituto de Estudios Superiores de la Empresa* [Institute of Higher Business Education] (IESE, per Spanish acronym), and other cases that can be classified as "semi-private philanthropic institutes."

The distribution of the population of centres by age is quite heterogeneous and, overall, it extends towards both extremes (Table 8.2). On one hand, there is a significant concentration of young centres: 51.9% of the centres were no more than 10 years old during the year in which we collected data (2012), meaning that their creation occurred after the year 2000. On the other hand, 19% of the population in 2012 were more than 20 years old, meaning that they were created or processed during the 1980s, or even earlier, during the period of transition to democracy. Perhaps the less dense area in the distribution of organisational age refers to the centres that are between 11 and 20 years old, overall, spanning between 15 and 20 years, meaning that the centres were created in the 1990s, which seems to be the period during which the creation of this type of organisation was less intense.

In conclusion, we observe a proliferation in the years of economic growth that may sometimes be interpreted as redundancy in the growth of the system due to the absence of structural changes in the traditional entities such as universities and Public Research Organisations (PROs). In this sense, future analysis of the pace at which CRCs were created during subsequent years up until this research (from 2012 forward) would be interesting, in order to find out if there had been significant changes during the second phase of the current economic crisis.

In Spain, CRCs are distributed across almost all 16 Autonomous Communities, with the exception of the Canary Islands and the autonomous cities of Ceuta and Melilla (Table 8.3). Andalusia is the Autonomous Community where most centres are located, with 36 centres making up 16.7% of the estimated population. Basque Country is in second place with 30 centres (13.9%) and in third, Catalonia, with 26 (12%); the Community of Valencia and Madrid share fourth place with 20 centres (9.3%) each. Cantabria and Navarre are the Autonomous Communities with the lowest number of CRCs, with three and five centres, respectively.

On one hand, the distribution of the centres among regions reflects the existing inequalities in terms of population (e.g. Andalusia occupies first place, Cantabria or Navarre in last) or R&D investment (e.g. the high score of Basque Country and

Table 8.3 Detailed Distribution of CRCs by Region

Region	N	%	No. of centres per 1000 R&D workers (FTE) in the region
Andalusia	36	16.7	1.52
Aragon	6	2.8	1.11
Asturias	11	5.1	3.53
Balearic Islands	8	3.7	4.33
Basque Country	30	13.9	1.68
Cantabria	3	1.4	1.68
Castile—La Mancha	7	3.2	0.79
Castile—Leon	7	3.2	2.49
Catalonia	26	12.0	0.59
Extremadura	6	2.8	3.15
Galicia	17	7.9	1.81
La Rioja	7	3.2	4.87
Madrid	20	9.3	0.43
Murcia	7	3.2	1.29
Navarre	5	2.3	1.13
Valencia	20	9.3	1.07
Total	216	100.0	1.08

Source: Map of centres (ES/CRC 2012); and Spanish statistical office (INE 2015); own elaboration

Madrid). On the other hand, we found some unexpected data, like the relatively high number of existing centres in Asturias (11) or the Balearic Islands (8), regions that are not exactly noted for their levels of R&D investment; or the comparatively low number that exist in regions in which R&D investment tends to be higher, such as Navarre or Castile and Leon. The communities that have the highest ratios in terms of personnel employed in R&D are La Rioja (4.87 centres per 1000 workers employed in R&D) and the Balearic Islands (4.33). Asturias (3.53), Extremadura (3.15) and Castile and Leon (2.49) fall behind those. However, the lowest ratios correspond to the Community of Madrid (0.43 centres per 1000 workers employed in R&D), Catalonia (0.59), and Castile-La Mancha (0.79).

If we group together all the regions in which the number of centres is below the average, we may be able to better observe the most relevant differences that exist between the regions as well as their relationship with the distribution according to type of CRC (Table 8.4). The relatively high and significant value of the association index indicates that the territorial distribution patterns vary significantly between the types of CRCs. Innovation and technology centres are essentially spread out evenly across Spanish geography, although a high number of ITCs stands out in Andalusia (20.9%) as well as the low number of ITCs located in the Community of Madrid (1.4%).

The *ad hoc* sites are distributed in a rather homogenous manner (Table 8.4), although they are somewhat common in the regions with fewer centres (38%) and in Madrid (16%). The centres in networks are almost exclusively located in Basque

Table 8.4 Regional distribution of types of CRC

	Taxonomy			
V = 0.401 (P value <0.001)	ITC (%)	Networks (%)	Ad hoc (%)	Total (%)
Andalusia	20.9		14.0	16.7
Basque Country	10.1	40.7	10.0	13.9
Catalonia	12.9	7.4	12.0	12.0
Galicia	10.1	3.7	4.0	7.9
Madrid	1.4	37.0	16.0	9.3
Valencia	11.5	3.7	6.0	9.3
Other	33.1		38.0	31.0
Total	100.0	100.0	100.0	100.0

Source: Map of centres (ES/CRC 2012); own elaboration

Country (40.7%) and Madrid (37%), plus a residual number in Catalonia, Galicia and Valencia, while they are not present in Andalusia or the rest of the regions.

8.4.3 *Knowledge Transfer Dynamics of CRCs*

Collaborative research centres facilitate a more proactive attitude about collaboration between scientific researchers and firms. This is based on the analysis of some indicators that come from the survey administered to the identified entities using the map of CRCs.[3] Below we present some results of the analysis that have to do with the dynamics of collaborative research between researchers and firms.

Figure 8.4 shows which type of entity the centre tends to partner with for each one of the activities that it performs. First, we see that CRCs tend to be organisations that collaborate rather frequently if we exclude firm creation, which is a less relevant activity. Collaboration in formalised activities is the most significant relationship, as reflected by the R&D projects and contracts. Second, we see that firms are the type of entity with whom centres collaborate most often, more specifically, in eight out of nine activities. For example, collaboration with firms is very important in contracted R&D projects and in services, while collaboration with universities or R&D organisations is greater in the case of R&D projects from public calls and in training of employees or doctoral students. Moreover, collaboration with public administration is residual in most cases, except in the case of publicly funded R&D projects.

Based on the data presented in Fig. 8.4, we constructed three variables that allowed us to synthetically measure the intensity of collaboration with each sector,

[3]The questionnaire was addressed to directors at the centres, or other administrative staff in charge. 128 centres participated in the survey, with a response rate of 59.3%. The analysis of some variables shows that there are no significant differences between the population and the sample (see also Fernández-Zubieta et al. 2016).

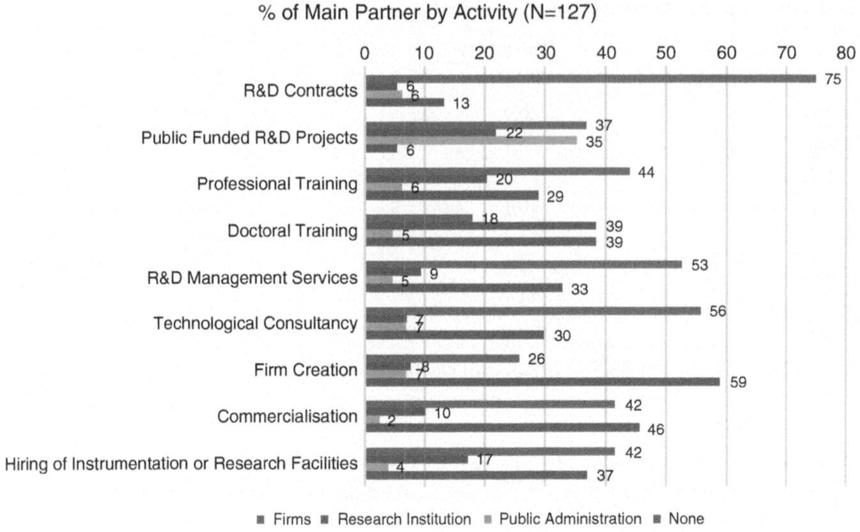

Fig. 8.4 Main partner by type of activity. Source: Survey to research centres (ES/CRC 2012); own elaboration

in terms of relevance.[4] We calculated three variables (one per sector) that assume the value of 0 if collaboration with that sector has no relevance, and 1 if it has maximum value. As expected, the variable that assumes on average the highest value is the one referring to collaboration with firms (Fig. 8.5), whose average is 0.306, followed by collaboration with R&D entities (average = 0.107) and public administration (average = 0.066).

The number of projects and activities performed is an important indicator that allows us to assess the orientation and volume of work from a more objective point of view. Table 8.5 shows the statistics related to the number of R&D projects and activities and services carried out by the centres in the last 3 years, differentiating between R&D projects and consultancy activities, technological assistance, and services contracted with administrations or firms. Among the projects, we differentiate between those contracted with administrations or firms, on one hand, and between those financed by national or international competitive calls, distinguishing them also upon whether the centre was the leader or member of the team. The most frequent activities are R&D projects from national public calls and those contracted with firms. The results show that during the last 3 years the surveyed centres have developed, in total, several thousands of R&D projects and activities, both in the public and private sector, as well as on an international level (Table 8.5). In the

[4]In order to construct these indicators, for each centre and for each sector we added up the number of activities where that sector was the main collaborator, weighing each unit by the relevance that the interviewed party attributed that activity. Then we divided this sum by nine, which is the total number of activities, and standardised the value.

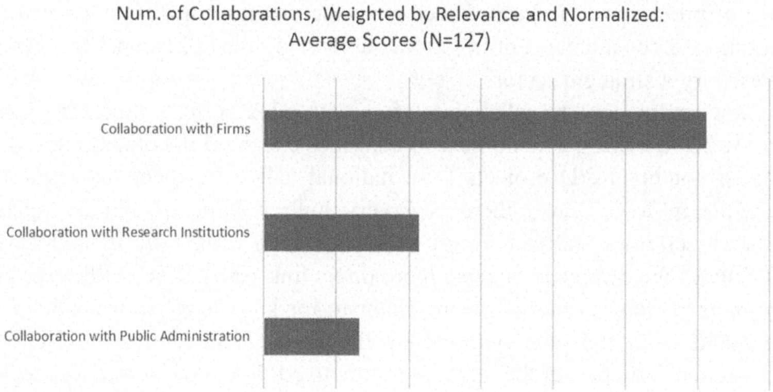

Fig. 8.5 Relevance of collaboration by sector. Source: Survey to research centres (ES/CRC, 2012); own elaboration

Table 8.5 Volume of collaborations by R&D projects and activities

Descriptive statistics (N = 122)	Public-funded R&D projects (Competitive calls)				R&D contracts		Consultancy, services	
	National		International		Pub. Adm.	Ind.	Pub. Adm.	Ind.
	Leader	Member	Leader	Member				
Mean	14.0	12.9	2.5	7.3	2.8	38.7	5.6	190.6
S.D.	21.8	27.5	6.1	20.4	9.0	103.4	21.8	690.2
Min	0	0	0	0	0	0	0	0
Max	120	200	50	200	79	800	175	5000
Sum	1703	1575	303	889	337	4723	686	23,248
1st Q	1.0	1.0	0.0	0.0	0.0	1.7	0.0	0.0
2nd Q	5.0	4.5	0.0	2.0	0.0	6.5	0.0	4.0
3rd Q	15.0	15.0	2.0	6.0	2.0	28.2	2.0	31.2

Source: Survey to research centres (ES/CRC 2012); own elaboration

last 3 years, each site has executed, on average: 13.9 R&D projects from national competitive calls as the leading centre and 12.9 as a member of the team; 2.5 R&D projects from international competitive calls as the leading centre and 7.3 as a member of the team; 2.8 R&D projects contracted with public administrations and 38.7 with firms; 21.8 consultancy activities and services with administrations and 190.6 with firms.

In summary, national policies guided towards the creation and diffusion of organisational innovations to favour science-industry linkage have been diverse, ranging from the creation of industrial associations, financing of collaborative projects, or diffusion of Technology Transfer Offices and Science and Technology Parks, to financing large public-private R&D consortia. However, it is on the regional scale that, recently, more innovative initiatives are taking place, like in the case of programs in industrial sectors or the creation of different networks of collaboration or consultancy entities to favour development of competitive synergies in the territory's strategic sectors.

The new centres tend to collaborate often with other entities, especially firms, to perform R&D activities. The number of activities executed in collaboration is quite high, most notably R&D projects from national calls or projects contracted with firms, while in fewer cases there is a very high volume of activity related to consultancy activities and technology services. These results are indirect proof of the capacity of the centres to create a more direct link between scientific researchers and firms, providing an innovative mechanism for knowledge transfer that can be distinguished from the plan outlined by the classical linear model. Hence, the companies that form part of the regional productive fabric have a source of technical scientific knowledge that can be taken advantage of more easily than the traditional forms of collaboration with universities and PROs. This ease would allow firms to adapt themselves more quickly to the changes that occur in the market or in the technological context during periods of crisis.

8.5 Conclusions

The results of this research shed light on data that help us to understand the relationship between the capacity of regional systems to create and adopt organisational innovations, and regional resilience. Without a sole frame of reference at the national level (as other countries do have), a core group of more proactive regions has constituted a group of different organisational innovations over the last 15 years, which have contributed to the heterogeneity of Spain's innovation system. We have seen how not all regions within the same country have the same capacity to innovate in science-industry relations despite sharing certain common problems related to the existing gap between science and industry, a strong dependency on productive sectors with low-intensity technology, and the assumption (sometimes implicit) of a linear model of knowledge transfer. Furthermore, this unequal capacity of the regions in creating organisational innovations does not necessarily seem to have a relationship with their size or competitiveness in the scientific or economic environment.

The results obtained reinforce the need to also consider the organisational dimension of innovation when studying the regional trajectories and dynamics. To that respect, there is evidence that organisational innovations that have been undertaken in Spain to strengthen the linkage between science and firms contributes to the resilience of innovation systems, at least in regard to the capacity for performing R&D. For example, the greater flexibility we see in the organisational structure of CRCs for carrying out collaborative research activities would imply that the territory's firms can more easily provide sources of technical scientific knowledge for solving urgent production problems, or even redesigning its production and commercialisation models using innovation technology solutions.

In summary, we highlight the relevance that organisational innovations have in the R&D systems of peripheral regions, as is the case in many Spanish regions. In these territories, the rigid bureaucratic structure, lack of policy planning, and existence of difficulties for collaboration act together as important barriers and have implications for regional resilience in peripheral regions. By acting upon these factors and reducing the distance between science and industry in social, cultural, and organisational terms, CRCs not only facilitate the strengthening of a territory's resilience and improvement of its capacity for adapting during periods of crisis, but they also contribute to unlocking the innovation trajectory of the peripheral regions.

Acknowledgments Financial support from the Spanish Plan for I + D + I of the Ministry of Economy and Competitiveness "New science-industry relationships: Cooperative Research Centres in the Spanish R&D System" (ref. cso2010 -14880. 2010-2013) is gratefully acknowledged. Sandro Giachi also acknowledges the financial support from the FPI *Formación del Personal Investigador* Programme (ref. BES-2011-047258).

References

Arias Aparicio, F. (2011). Los centros de investigación: variedad tipológica y diversidad jurídica. In L. S. Menéndez & L. C. Castro (Eds.), *La investigación y sus actores. Institutos y centros de I+D y sus desafíos*. Barcelona: Fundación CYD.

Boschma, R. (2015). Towards an evolutionary perspective on regional resilience. *Regional Studies, 49*(5), 733–751.

Christopherson, S., Michie, J., & Tyler, P. (2010). Regional resilience: Theoretical and empirical perspetives. *Cambridge Journal of Regions, Economy and Society, 1*, 3–10.

Clark, J. (2010). Coordinating a conscious geography: The role of research centers in multi-scalar innovation policy and economic development in the US and Canada. *The Journal of Technology Transfer, 35*(5), 460–474.

Clark, J. (2015). *Informe COTEC 2015: tecnología e innovación en España*. Madrid: Fundación COTEC para la innovación tecnológica.

COTEC. (2007). *Colaboración público-privada en innovación (Encuentros empresariales n. 14)*. Madrid: Fundación COTEC para la innovación tecnológica.

Cruz-Castro, L., & Sanz-Menéndez, L. (2007). New legitimation models and the transformation of the public research organizational field. *International Studies of Management & Organization, 37*(1), 27–52.

Díaz-Catalán, C., Hernández-Hernández, N., & López-Navarro, I. (2011, September 15). *Evaluation system barriers for the third mission in the universities*. Atlanta conference on science and innovation policy. Atlanta.

European Commission. (2010a). *Europe 2020. A European strategy for smart, sustainable and inclusive growth*. Luxembourg: Publications Office of the European Union.

European Commission. (2010b). *Europe 2020 flagship initiative innovation union*. Luxembourg: Publications Office of the European Union.

FECYT. (2015). *Indicadores del sistema español de ciencia, tecnología e innovación*. Madrid: Fundación Española para la Ciencia y la Tecnología.

Fernández-Esquinas, M., & Ramos-Vielba, I. (2011). Emerging forms of cross-sector collaboration in the Spanish innovation system. *Science and Public Policy, 38*(2), 135–146.

Fernández-Esquinas, M., Giachi, S., & Pérez-Yruela, M. (2012, August 8–10). *Building a triple helix corporation: Organizational innovation for cross sector research collaboration in a catch-up region*. 10th Triple Helix International Conference. Bandung (Indonesia).
Fernández-Zubieta, A., Andújar-Nagore, I., Giachi, S., & Fernández-Esquinas, M. (2016). New organizational arrangements for public-private research collaboration. *Journal of the Knowledge Economy, 7*(1), 80–103.
Garrett-Jones, S. (2007). Knowledge and cooperation for regional development: The effect of provincial and federal policy initiatives in Canada and Australia. *Prometheus, 25*(1), 31–50.
Gómez Gras, J. M., Solves, I. M., & Jover, A. J. V. (2007). Las spin-off académicas como vía de transferencia tecnológica. *Economía Industrial, 366*, 61–72.
Gray, D. O., Boardman, C., & Rivers, D. (2013). The new science and engineering management: Cooperative research centers as intermediary organizations for government policies and industry strategies. In C. Boardman, D. O. Gray, & D. Rivers (Eds.), *Cooperative research centers and technical innovation: Government policies, industry strategies, and organizational dynamics* (pp. 3–33). New York: Springer.
Hage, J. T. (1999). Organizational innovation and organizational change. *Annual review of sociology, 25*(1), 597–622.
Iglesias Sánchez, P., Maldonado, C. J., & Velasco, A. P. (2014). Comparación de la actividad de I+ D+ i de Spin-Off universitarias y otras organizaciones empresariales. *Economía industrial, 392*, 155–168.
Lam, A. (2005). Organizational innovation. In J. Fagerberg, D. Mowery, & R. R. Nelson (Eds.), *The Oxford handbook of innovation* (pp. 115–147). Oxford: Oxford University Press.
Liyanage, S. (1995). Breeding innovation clusters through collaborative research networks. *Technovation, 15*(9), 553–567.
Manjarrés-Henríquez, L., Gutiérrez-Gracia, A., Carrión-García, A., & Vega-Jurado, J. (2009). The effects of university–industry relationships and academic research on scientific performance: Synergy or substitution? *Research in Higher Education, 50*(8), 795–811.
Martin, R., & Sunley, P. (2015). On the notion of regional economic resilience: Conceptualization and explanation. *Journal of Economic Geography, 15*(1), 1–42.
Mulet Meliá, J. (2003). Una visión empresarial del sistema español de innovación. *Economía industrial, 354*, 25–36.
Palomares Montero, D., García Aracil, A., & Castro Martínez, E. (2012). Misiones actuales de las universidades públicas: una perspectiva sociológica. *Arbor, 188*(753), 171–192.
Perkmann, M., & Walsh, K. (2007). University–industry relationships and open innovation: Towards a research agenda. *International Journal of Management Reviews, 9*(4), 259–280.
Pinto, H. (2015). *Resilient territories: Innovation and creativity for new modes of regional development*. Cambridge: Cambridge Scholars Publishing.
Pinto, H., & Pereira, T. S. (2013). Efficiency of innovation systems in Europe: An institutional approach to the diversity of national profiles. *European Planning Studies, 31*(6), 755–779.
Sanz-Menéndez, L., & Cruz-Castro, E. (2005). Explaining the science and technology policies of regional governments. *Regional Studies, 39*(7), 939–954.
Simmie, J. (2014). Regional economic resilience: A schumpeterian perspective. *Raumforsch Raumordnung, 72*, 103–116.
Simmie, J., & Martin, R. (2010). The economic resilience of regions: Towards an evolutionary approach. *Cambridge Journal of Regions, Economy and Society, 3*, 27–43.
Tortosa, E. (2006). La I+ D en el marco autonómico. In J. Sebastián & E. Muñoz (Eds.), *Radiografía de la investigación pública en España* (pp. 71–95). Madrid: Biblioteca Nueva.
Vence, J., & Heijs, J. (2006). Estructura y flujos de la financiación de la I+ D en España. In J. Sebastián & E. Muñoz (Eds.), *Radiografía de la investigación pública en España* (pp. 209–272). Madrid: Biblioteca Nueva.

Chapter 9
Merging Entropy in Self-Organisation: A Geographical Approach

Eric Vaz and Dragos Bandur

9.1 Introduction

The impact of scale and aggregation associated with administrative boundaries on spatial data modeling has been studied extensively (Vaz et al. 2015). For example, Openshaw (1977) provides a case study involving changes in linear correlation of early- and mid-Victorian house characteristics as the grid cell size in which the study area was partitioned increases gradually from 100 m to 1 km. Demonstrated that, for the same dataset, there are multiple ways of aggregation which could render a correlation strength anywhere between −1 and +1, and that results differ significantly between different aggregation schemes. From a spatial perspective, when the partition units are irregular in shape and area, the number of ways of partitioning the same region, given the scale, becomes practically unlimited, increasing the complexity of the problem further. In this respect, existing and fixed administrative divisions could be regarded as realizations of a partitioning process and, as suggested in a more recent health study on Ontario's population—which places the Modifiable Areal Unit Problem (MAUP) at the center of disagreements concerning many quantitative regional studies—the "a priori identification of the scale and area underlying each study" becomes a necessity (Parenteau and Sawada 2011).

A similar scale and aggregation problem face health information data which, motivated by the preservation of patient's privacy, among other reasons, are mostly available in an aggregated form that usually corresponds to certain levels of administrative boundaries. To complicate the issue further, administrative divisions (e.g. Census Tracts, Census Divisions, etc.) do not match the Health Divisions (e.g. Public Health Regions, Local Health Integration Networks, etc.) and, as result,

E. Vaz (✉) · D. Bandur
Department of Geography and Environmental Studies, Ryerson University, Toronto, ON, Canada
e-mail: evaz@ryerson.ca

H. Pinto et al. (eds.), *Resilience and Regional Dynamics*, Advances in Spatial Science, https://doi.org/10.1007/978-3-319-95135-5_9

health information data related to the former must undergo further transformations before being reported in relation to the later, which are complicated by the many-to-many relationships between the partition schemes. Given the arbitrary character of the administrative divisions in relation to spatial data under investigation, the relationship between aggregation, data complexity and access to information remains partial and ambiguous (Openshaw and Rao 1995). There is however, a less arbitrary form of aggregation since the driver for it is within the information patterns and the intrinsic segregation of the data. Such aggregation types emerge as the natural choice in classification exercises, accomplishing at least two tasks: the reduction of data complexity by increasing its level of aggregation and creating information content more (if not completely) accessible through rearranging the data by an extended form of similarity (e.g. Mutual Information). Since data represent records of real-world events, the spatial variables implicated in this project are considered regionalized variables, i.e. realisations of underlying stochastic phenomena. Consequently, the term "entropy" refers to the probabilistic definition given by Shannon (2001) to information entropy in the context of Information Theory (Cover and Thomas 2006). In this respect, the entropy is introduced as the average uncertainty (in the sense of statistical expectation) contained in one given variable. Rethinking spatial aggregation brings forth two questions that this chapter attempts to answer: Does data re-organisation reveal more of information in the data? When associated with data, is entropy a tool to sensitively and consistently measure such changes? By using entropy as a "criterion for testing hypotheses about systematic effects in experiments where frequency data are available" (Jaynes 1982), this chapter responds to the evaluating the changes in entropy and information driven by sequential data aggregation through the Self-organising Maps (SOM) algorithm. This will be achieved by a series of integrated steps: (i) evaluation of entropy associated with the input data, (ii) realization of increasing levels of spatial aggregation (decreasing complexity) as result of an iterative application of the SOM algorithm, (iii) evaluation of entropy, mutual information and spatial mutual information at each level of spatial aggregation (Batty 1974), and (iv) observing the empirical relationship between information and spatial aggregation.

9.1.1 Entropy, Spatial Entropy and Mutual Information

In classical physics, the entropy was introduced as an *extensive* thermodynamic parameter to select among the thermodynamic processes those with a natural direction of evolution involving an ever increasing entropy. This parameter was introduced through the Second Principle of Thermodynamics which states, in essence, that thermodynamic process evolves in the direction of increasing or, of stationary entropy. This principle is the second thermodynamic principle set to introduce a non-mechanical concept to the classical physics; the other principle (called The Principle Zero of Thermodynamics) introduces the notion of temperature as an *intensive* thermodynamic parameter, postulating the equality of temperatures as condition for thermal equilibrium between two systems in mechanical contact. As result, classical thermodynamics employs mechanical parameters such as internal energy and mechanical work and non-mechanical parameters such as temperature and entropy to describe the thermodynamic evolution of systems. While the presence and significance of mechanical parameters was obvious, there was no explanation for temperature and entropy that could satisfy the determinism of classical mechanics. Therefore, with the development of Kinetic Theory of Gases, entropy re-surfaced as a measure of probability with which a system reaches a certain state.

This statistical view, attributed to the Austrian physicist and philosopher Ludwig Boltzmann, initially faced a stern resistance in the academic community. The formula advanced by Boltzmann was as follows:

$$S_B = -k_B \ln(\Omega) \tag{9.1}$$

where S_B represents Boltzmann entropy, k_B represents the ratio between particle's energy and the equilibrium temperature of the system containing the particle (heat capacity), Ω represents the number of all possible micro states compatible with a certain macro state of a system. The evolution of Boltzmann's entropy gained a larger interest within the social sciences through its application by Claude Shannon. In his derivations, Shannon established that, *if* there was a measure for an outcome's uncertainty in a sequence of random events of probabilities (p_i), *then*: (1) the measure would be continuous in these probabilities, (2) it would reach the maximum value when all probabilities are equal (i.e. equally likely events are harder to predict) and, (3) should a partitioning of this sequence be possible, the measure would be the weighted sum of its components derived from each partition. According to the Principle of Maximum Entropy formulated by Jaynes (1982), *"when our a priori information is incomplete, our inferences should be drawn from the probability distribution that has the maximum entropy permitted by the information we do have"*—this probability distribution is the most unbiased and conducive to new information. In one of his articles, Jaynes worked to remove vagueness from terms and definitions surrounding Shannon's entropy by asking: "just how strongly [are the] distributions of lower entropy ruled out?". Entropy of a single random variable is regarded equally as uncertainty or information contained within the variable. Once the number of random variables increases, this equivalency is lifted. In general, multiple random variables (X, Y, ...), which may constitute the components of a random vector, have an associated entropy defined by joint probabilities.

If entropy is regarded as the average amount of information required for describing a distribution {p} then, it is expected that, by wrongly assuming a different distribution {q}, a different amount of information would be required, and this amount could only be larger than the amount of information required by the "right" distribution {p}. An efficiency measure for this difference is known as the Kullback-Leibler (K-L) "distance" between distributions.

K-L Distance introduces the mutual information between two random variables in terms of distance between the joint and independent states of these variables as follows:

$$I(X,Y) = D(p(x,y)\|p(x)p(y)) = \sum\sum p(x,y)\log(p(x,y)/p(x)p(y)) \quad (9.2)$$

where p(x) and p(y) are marginal probabilities of random variables X and Y, and p(x,y) represents their joint probability. As such, mutual information *I*(X,Y) represents the "distance" between the joint distribution of two random variables and their distribution at maximum entropy case in which, distributions {p} and {q} are independent according to the Principle of Maximum Entropy. Mutual information is also interpreted as the reduction of uncertainty in one random variable as result of the knowledge of the other (Cover and Thomas 2006).

9.1.2 *Spatial Entropy*

Attempts at reconciling Shannon's information entropy for discrete and continuous probability distributions led to a modified formulation for the entropy known as Spatial Entropy (Batty 1974). In this sense, the reconciliation requires a spatial element resulted from the partitioning of spatial support (the data domain), which brings into light: (i) an associated measure for the spatial extent and (ii) a functional reference to a geometric or a geographic coordinates system. As such, starting with the aspatial formulation of entropy for discrete case, with probabilities p_i written as functions of probability densities $p(x_i)$, one obtains:

$$p_i = p(x_i)\Delta x_i \text{ and} \sum p_i = 1 \quad (9.3)$$

The entropy would arguably increase indeterminately as $\Delta x_i \rightarrow 0$, due to the implied proportionality between probability and the extent of the spatial support, which leaves Spatial Entropy as:

$$S = -\sum p_i \log(p_i/\Delta x_i) \quad (9.4)$$

9.1.3 *Self-Organising Maps*

Self-organising Maps (SOM) are a special class of ANN that are capable of unsupervised learning and therefore adept of performing ad-hoc (unsupervised) classifications. Since their introduction in early 1980s by Kohonen, the use of SOM algorithm has expanded continuously in the fields of technology and research, from environmental analysis, text mining and image processing, to fraud detection and manufacturing process control. In addition to their short execution time of large datasets, data clustering capabilities, etc., SOM are an excellent tool for data visualisation. All of these characteristics, as well as the increasing general interest in their application, motivated the decision to select the SOM algorithm as a data aggregation tool for this project.

When applied to spatial data, SOM algorithm generates two types of maps: (1) the neural grid, which is usually a 2-dimensional projection (or abstraction) of the multi-dimensional input data cloud, and (2) the geographic (or geometric) representation of the output, re-organised (clustered, classified and projected) according to the algorithm's parameter settings. Application of SOM to spatial data gives two types of spatial reference: (1) a reference to the neural grid with the help of synaptic weight vectors, and (2) a geographic or projected coordinate system shared between output and input. The result is a sequential re-organisation of the study area through changes in aggregation levels.

Neurons that end up as the centroids of input data clusters (as result of a competitive/collaborative learning process) are called "winning neurons". They form a subset of SOM grid neurons. During the learning process, each data record is indexed and becomes member to a group of similar records, which define the clusters. Aggregated groups are built by similarities, records belonging to one group are similar in some respect, and dissimilar from records belonging to other groups. Although all input records are indexed, records belonging to the same group receive a similar index value. At low levels of aggregation, there may be less data segments than SOM grid neurons.

9.2 Data

The datasets used in our analysis have been obtained from the Ontario Ministry of Health and Long Term Care (MoHLTC) and contain case records of addiction and mental health aggregated to Census Division level (Fig. 9.1).

The data, initially aggregated to Census Division level, contain addiction and mental health population information for the year 2011. It was collected for the purpose of a Practicum project, which established new locations throughout the Province of Ontario for the Ontario's Centre of Addiction and Mental Health Service Collaborative Teams, and the data included recorded cases between 18 years and 64 years of age. Table 9.1 provides a snapshot of this data.

The file contains 13 variables and 49 rows of recorded counts. Variable names were coded as follows:

1. linc = Individual low-income status (after tax)
2. aborig = Aboriginal
3. french = Francophone population (defined as language spoken at home)
4. nofreng = Population who speak no official language (English or French)
5. oadcomh = Cases with open admissions to substance abuse programs who report concurrent mental health problems

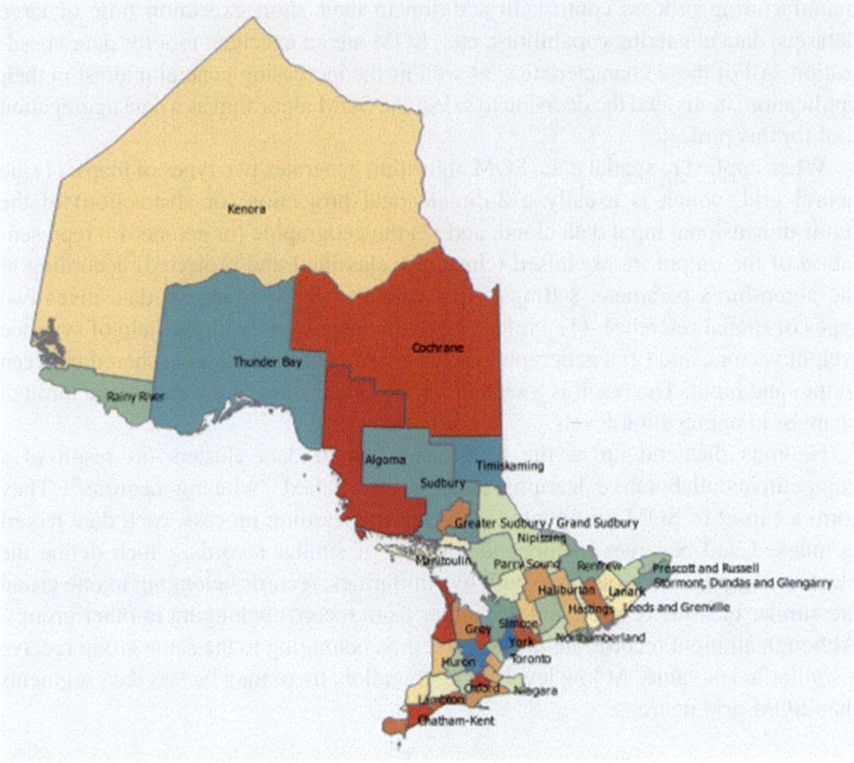

Fig. 9.1 Census divisions within the province of Ontario

Table 9.1 Counts of cases

Coded Variable Names	*oadcomh*	*pd30d*	*a30d*	*na30d*	*aborig*	*nofreng*
Census division	795	222	32	26	2340	14,410
Stormont, Dundas and glengarry	463	193	35	22	1470	48,535
Prescott and Russell	4081	1900	313	214	12,250	84,970
Leeds and Grenville	448	136	41	29	1760	815
Lanark	393	133	29	17	1575	700
Frontenac	1103	464	129	93	3360	2005

6. pd30d = Cases with 30-day post-discharge mental health and addiction OHIP visit
7. na30d = Cases with 30-day post-discharge mental health and addiction ED visits but not admitted (18 to 64 years old)
8. a30d = Cases with 30-day mental health and addiction readmission (18 to 64 years old)
9. navisit = Count of mental health and addiction/Emergency visits where the referral source is police/corrections and disposition is NO ADMISSION

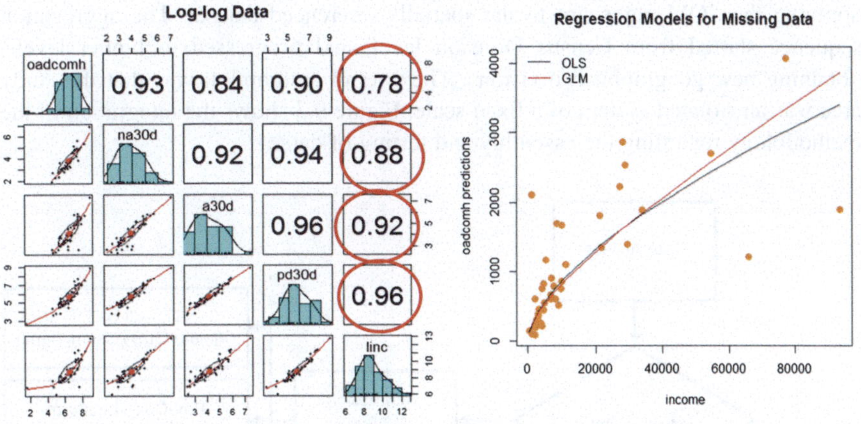

a) b)

Fig. 9.2 (**a**, **b**) Example of missing data replacement with linear regression fitted values

10. code_1 = Count of self—perceived fair or poor mental health
11. code_3 = Count of current smoker, smokes daily
12. code_4 = Count of second-hand smoking exposure
13. code_5 = Count of heavy drinkers

Missing data were replaced with log-linear regression fitted values using the "income" as predictor variable. Fig. 9.2b provides a visual comparison between Ordinary Least Squares and Generalised Linear Model predictions for the variable number 5 in the list above.

Sample joint frequencies associated with recorded cases observed in each division unit were calculated as ratios between case counts and divisional population counts. Ratios were used because they constitute the maximum likelihood estimates for joint frequencies and, for the purpose of the study, all measures of entropy were estimated using James-Stein type shrinkage estimators, well-suited for small samples (Hausser and Strimmer 2009). Appendix E contains a short presentation of these estimators.

9.3 Methodology

First, the entropy of the SOM grid for various grid sizes was calculated and then the mutual information of the spatially-referenced health information dataset, which followed the aggregation levels that resulted from the SOM algorithm. The values obtained for various levels of aggregation corresponding to various SOM grid sizes were plotted and discussed. Following the Principle of Hierarchical Decomposition of Entropy, an empirical relationship between entropy and mutual information on one side, and the level of data aggregation on the other, was expected as result of applying the SOM algorithm to the spatially referenced dataset. The aggregation sequence started from Census Division level, and progressed to higher levels, obtaining new geographical divisions. The overall assumption was that the study area was an isolated system of a fixed scale. Figure 9.3 shows the integration of the methodology regarding the assembly and testing of data.

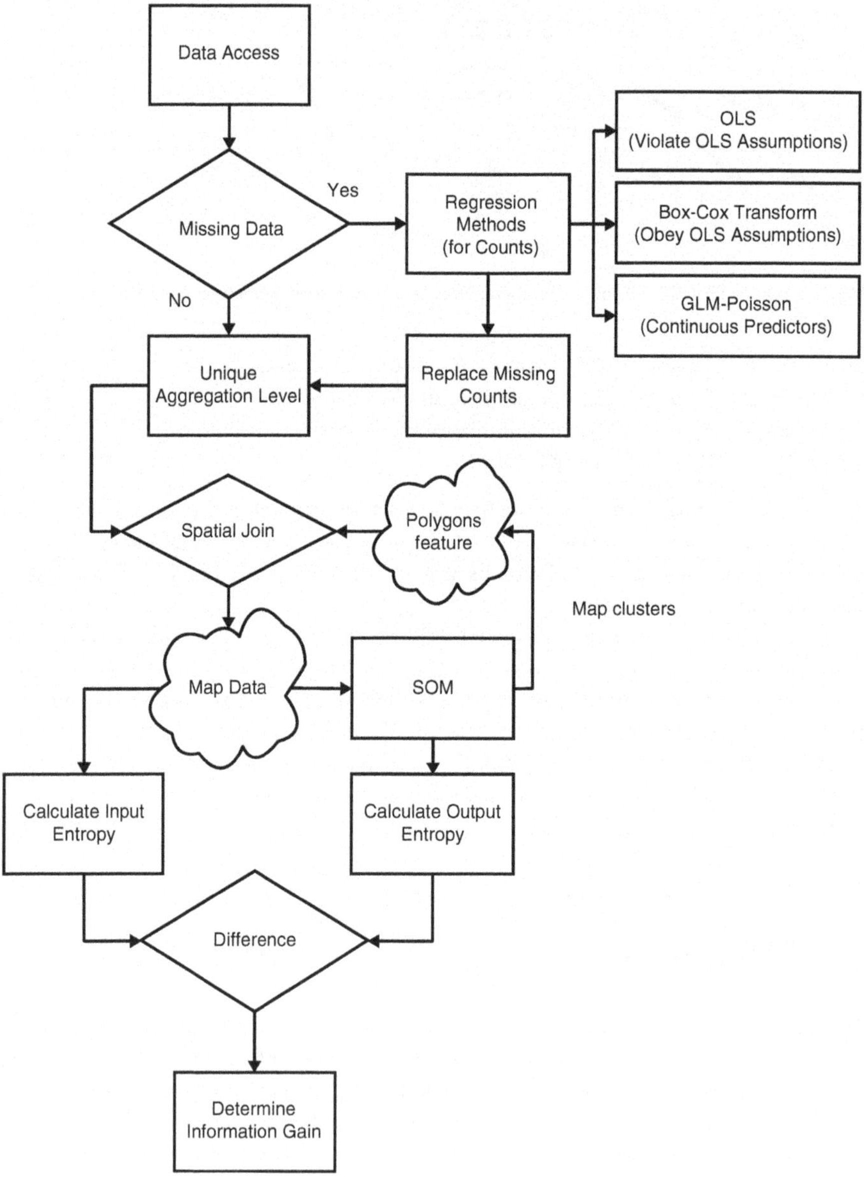

Fig. 9.3 Methodology

9.4 Results

The analysis closes with graphic presentations of the statistically significant correlations between geographical divisions using mutual information network graphs for three levels of aggregation. Since the formulation of spatial mutual information uses probabilities weighted by the importance of spatial divisions, it should exhibit a strong correlation between variables X and Y defined above, at least up to a certain level of spatial aggregation. The expectation was that the underlying correlation would be stronger than in the aspatial version.

The exercise was conducted under the assumption that the study area was an isolated system of a fixed scale and that the type and level of aggregation were the only modifications taking place, as accomplished through iterative application of a SOM algorithm of variable neural grid size. Grid entropy as percent of maximum entropy by grid size (defined as the number of neurons in the grid) is derived and presented initially using the maximum likelihood-estimated (MLE) frequencies. Aspatial and spatial mutual information associated with the dataset are derived through iterative spatial aggregation of the data, providing a probabilistic link between the attribute and spatial characteristics of the data.

Shannon's information entropy and relative entropy were calculated for all height-width grid combination pairs, up to a size of 48 grid cells (neurons). Appendix F contains examples of R scripts used in analysis.

A breaking point at 14 grid neurons delimitates the case in which, for this dataset, the numbers are equal, suggesting at least as many clusters in the dataset as there are neurons in the grid. Above this breaking point there are more grid neurons than data clusters—all neurons are therefore not associated with data clusters (Table 9.2).

Figure 9.4 depicts a number of height-width grid configuration pairs containing equal number of neurons but different number of winners. The most obvious is the case of 42-neuron grid size, with the corresponding pairs $i = 6, j = 7$ and $i = 7, j = 6$. The totals of winning neurons for these pairs are 30 and 27, respectively. This difference in winners corresponds to different count distributions across the grid and to different entropy and relative entropy values and count distributions, implicitly.

The largest observed K-L Distance roughly corresponds to the 10-neuron grid size, which has an equal number of winners and neurons according to Table 9.3.

The count distribution corresponding to this configuration is shown in Fig. 9.5 as not uniform. Transmission of information through any ANN is triggered by neuron activation; therefore, the state of maximum entropy corresponds to an equal activation probability for all the neurons, independent of the "semantic structure of the transmitted signal" (Polani 2002). The entropy changes depicted in Fig. 9.5 are therefore only apparent as they are driven by the drop in entropy maxima, which depends of the

Table 9.2 SOM grid entropy and Kulback-Leibler distance associated with SOM grid size (sample)

I	J	H	K-L Dist.	Hmax	Winners	Tot. neurons
1	2	0.1437262	0.8562738	1.000000	2	2
1	3	0.8257300	0.7592325	1.584963	3	3
2	2	0.9181652	1.0818348	2.000000	4	4
1	4	0.9181652	1.0818348	2.000000	4	4
1	5	1.1829518	1.1389763	2.321928	5	5
⋮						
2	4	1.7692813	1.2307187	3.000000	8	8
3	3	1.9598147	1.2101103	3.169925	9	9
5	3	2.7940398	1.1128508	3.906891	14	15
⋮						
7	6	4.3640116	1.0283058	5.392317	27	42
6	7	4.5788958	0.8134216	5.392317	30	42
7	7	4.6276787	0.9870312	5.614710	30	49

i and j identify SOM neural grid row respectively, column
$1 \leq i, j \leq$ Number-of-rows in the dataset
H represents the entropy of the SOM grid
Hmax represents maximum entropy for each grid size
K-L Distance represents the relative entropy (K-L Distance) between entropy and its maxima

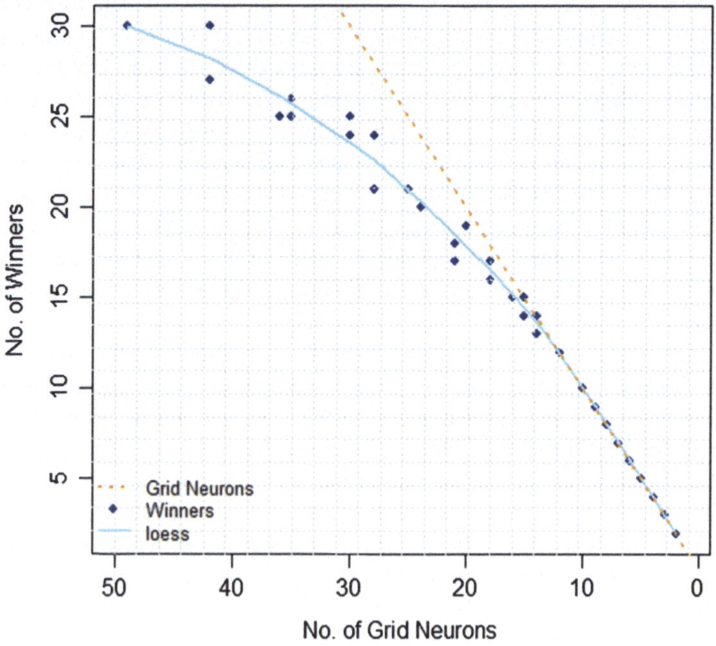

Fig. 9.4 The balance between grid neurons and winning neurons by SOM grid size

Table 9.3 Entropy maximum entropy and relative entropy for a 10-neuron SOM grid size

I	J	H	KL_D	H Max.	Winners	Grid neurons
5	2	2.016	1.305	3.321	10	10
2	5	2.437	0.884	3.321	10	10

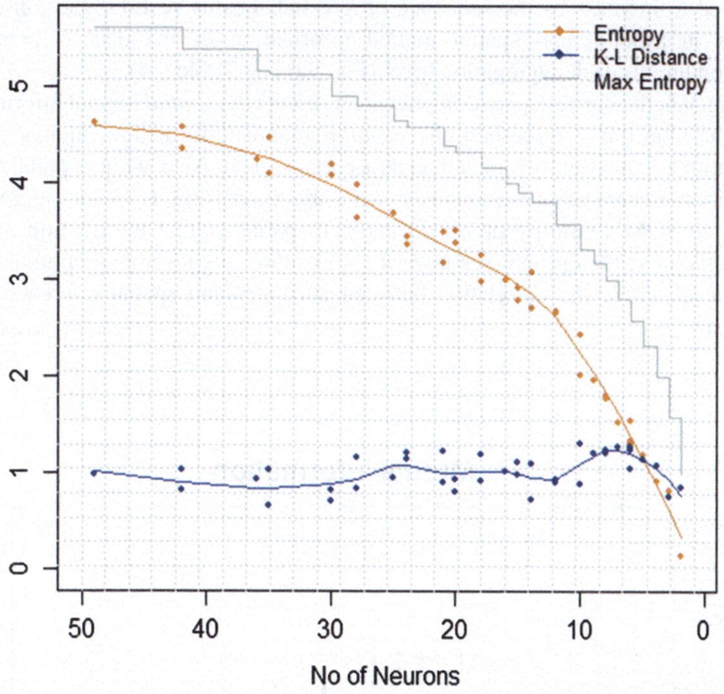

Fig. 9.5 The difference (in blue) between grid entropy and maximum grid entropy for each grid size

drop in grid size only. This suggests that, in some cases, the entropy changes may be unrelated to the data and consequently entropy requires normalisation.

The relatively unchanged K-L Distance between the 48-neuron and 25-neuron grid sizes suggests that the entropic drop (orange curve) is driven by the drop in entropic maxima (gray lines) which, in turn, are controlled by the reduction in grid size. Figure 9.5 shows an *apparent* entropic drop above the 12-neuron grid and a *real* drop between the 12-neuron and 2-neuron grid sizes, where K-L Distance goes through a maximum value, point where the system is in the most informative state, according to Batty and eq. (13).

The partitioning of the Province of Ontario corresponding to 10-neuron grid is depicted where newly-created divisions are identified by numbers. The new divisions are contiguous (e.g. division number 8), as well as non-contiguous (e.g. division numbers 1, 2, 3, 9).

9.4.1 Aspatial and Spatial Mutual Information

Spatial aggregation of the data was driven by a sequential change in SOM grid size presented previously. The aggregation started from the Census Division level and continued to increase as the sequence progressed. In this section, the consequent changes in aspatial and spatial mutual information of the data is presented, corresponding to each aggregation level (Fig. 9.6 and Table 9.4).

In Fig. 9.6, the curve depicted in orange represents the *spatial* mutual information, while the blue curve represents the *aspatial* mutual information. Spatial mutual information is consistently higher than its aspatial variant. Also, while aspatial mutual information decreases monotonously with the increasing level of aggregation (similar to SOM's entropy presented earlier), spatial mutual information remains quasi-stationary, then progresses toward a maximum, suggesting an optimal aggregation level where the correlation between attributes and spatial reference is the strongest.

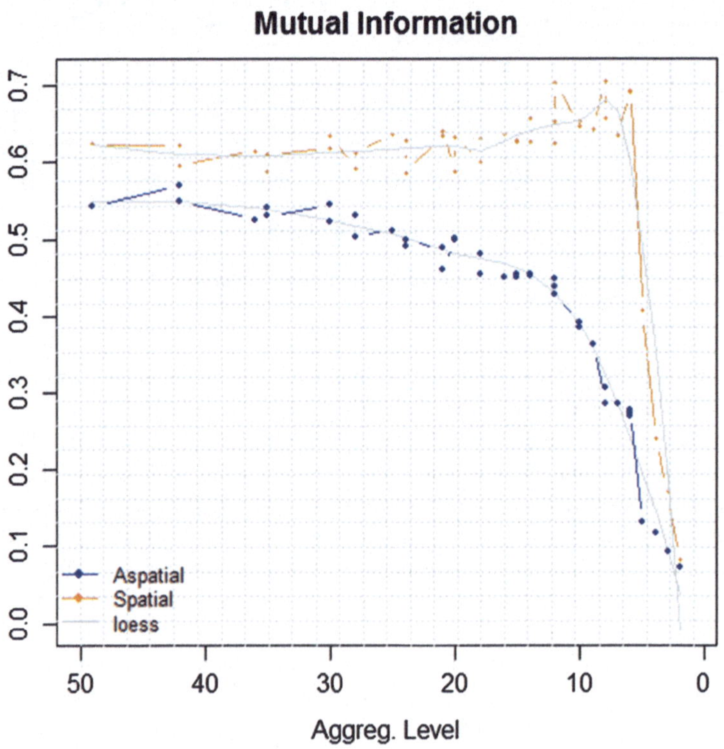

Fig. 9.6 Difference between aspatial and spatial mutual information values calculated from Ontario MoHLTC dataset

Table 9.4 Mutual information associated with spatial aggregation level of data (sample)

I	J	Aspatial	Spatial	Aggreg. level
1	2	0.07297769	0.08189621	2
1	3	0.09242060	0.17187349	3
2	2	0.11779010	0.24090009	4
⋮				
2	3	0.27000006	0.69270229	6
1	6	0.27399767	0.69450908	6
1	7	0.28546622	0.63480756	7
4	2	0.30514440	0.65734147	8
2	4	0.28655320	0.70600448	8
3	3	0.36347608	0.64306875	9
5	2	0.38532311	0.65370352	10
2	5	0.39097016	0.64783732	10
⋮				
6	6	0.52630407	0.61585686	36
7	6	0.54999902	0.59637495	42
6	7	0.56921135	0.62234137	42
7	7	0.54422934	0.62423604	49

Table 9.5 Aspatial and spatial mutual information maxima

I	J	Aspatial	Spatial	Aggreg. level
2	4	0.287	0.706	8
4	2	0.305	0.657	8

Columns 3 and 4 of Table 9.4 contain mutual information values. Column 5 contains the number of SOM neurons in inverse relation with the level of spatial aggregation.

The maximum value of spatial mutual information was found through LOESS smoothing, corresponding to a grid size of 8 neurons (Table 9.5):

These results suggest that:

- Data-oriented aggregation (based on intrinsic data segmentation) could prove to be a more natural solution to aggregation than aggregation based on administrative units, as it insures and maintains a strong correlation (in the extended meaning of mutual information) between attributes data and their geographical reference and extent;
- An optimal partitioning configuration corresponding to a certain aggregation level is possible in which, spatial mutual information is maximum. The (non-uniform) distribution of the recorded health cases, corresponding to the optimal grid size, and the associated geographic partition.

The similarity between the patterns of relative grid entropy and spatial mutual information are shown in Fig. 9.7 below. For both measures, the observed maxima appear between 5 and 10 grid neurons. LOESS smoothing places them both at 8-neuron grid size.

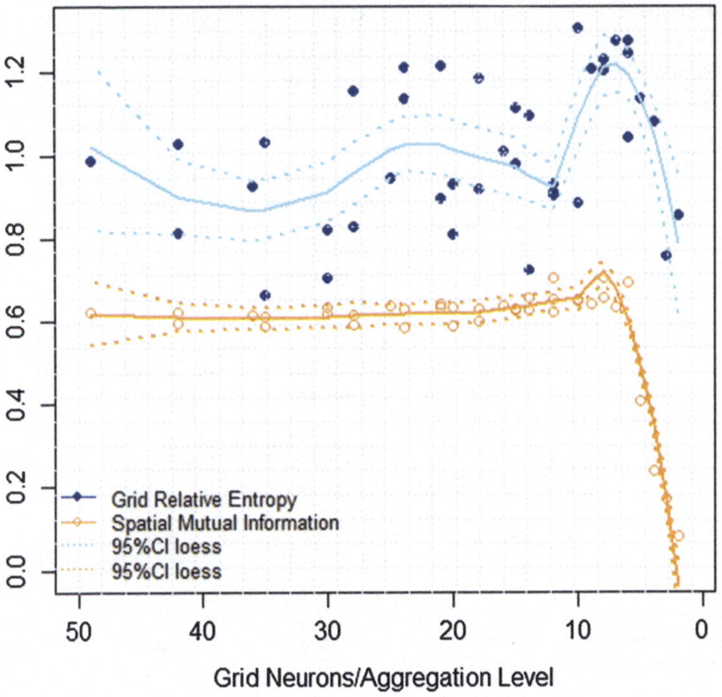

Fig. 9.7 Grid relative entropy and spatial mutual information by aggregation level

9.5 Conclusions

In conclusion, SOM grid relative entropy and spatial mutual information associated with the data demonstrate the adaptive character of SOM-driven data aggregation. Through this type of aggregation, mutual information is preserved as long as the importance (weight) related to the spatial extent of each division is considered. The Information Network is a highly informative tool which helps visualizing hierarchical relationships that take place within complex systems.

The information within a certain dataset builds upon many factors, including collection, pre-processing, data type, subject matter, spatiality, etc. These factors affect the internal structure and segmentation of what we ultimately refer as "the data". Consequently, a data-oriented aggregation method that closely follows patterns could yield different partitions of the same study area and aggregation level for different datasets.

Further, three mutual information networks plotted for the 49, 10 and 8-neuron grid sizes, respectively, corresponding to increasing aggregation level of data; each network describes a certain aggregation level. The vertices symbolize 95% Significance Level-significant correlations between the nodes representing geographic divisions that have been obtained through data aggregation.

With the increased level of aggregation (Figs. 9.3 and 9.6), the system's complexity diminishes. At Census Division level, the divisions form complex connections with isolated clusters and super-clusters containing parent nodes strongly correlated with neighbors, as specified by Tobler's Law of Geography. As the level of aggregation increases, nearest neighbors coalesce and super-clusters shrink, leaving room for simpler networks until the complete disappearance of hierarchy. At the highest level of aggregation, no parent nodes remain and only 5 out of the 8 existing divisions maintain significant levels of correlation.

Two novel findings that contribute to spatial analysis were registered: (i) SOM grid entropy and aspatial mutual information decrease monotonously as the SOM grid size decreases and the level of aggregation in the data increases correspondingly; and (ii) SOM grid relative entropy (K-L Distance) and spatial mutual information of data do not behave monotonously as they show simultaneous maxima at a certain level of spatial aggregation. Throughout this study, it has been observed that a decreasing complexity associates with a monotonously decreasing *aspatial* mutual information, which ultimately reaches a complete disconnect between the attribute and their spatial reference, a situation corresponding to a maximum entropy state also found within spatial interaction (Hagen-Zanker and Jin 2012).

Mutual information weighted by spatial extent remains preserved throughout the aggregation, anticipating a general data-oriented method for data aggregation without loss or distortion of information and should be taken into account when addressing regional phenomena. Ultimately, spatial mutual information also reaches the maximum entropy state, but it does so after passing through a maximum associated with an "optimal" aggregation level and a corresponding partition where attribute and spatial reference correlate strongly, and where both data and SOM grid find themselves in a highly informative state. In this sense, regional boundaries should entail a framework in both self-organization and a data-explicit context, so as to find the ideal topology for the most adequate representation of regional phenomena linked to spatial change and distribution. Ultimately, this will contribute to encouraging the construct of regional intelligence (Vaz 2018) where multiple regional scales can thus be applied for a given spatial extent, independent of administrative boundaries and configuration.

References

Batty, M. (1974). Spatial entropy. *Geographical Analysis, 6*(1), 1–31.

Cover, T. M., & Thomas, J. A. (2006). *Elements of information theory* (2nd ed.). Hoboken, NJ: Wiley.

Hagen-Zanker, A., & Jin, Y. (2012). A new method of adaptive zoning for spatial interaction models. *Geographical Analysis, 44*(4), 281–301.

Hausser, J., & Strimmer, K. (2009). Entropy inference and the James-stein estimator, with application to nonlinear gene association networks. *The Journal of Machine Learning Research, 10*, 1469–1484.

Jaynes, E. T. (1982). On the rationale of maximum-entropy methods. *Proceedings of the IEEE, 70*(9), 939–952.

Openshaw, S. (1977). A geographical solution to scale and aggregation problems in region-building, partitioning and spatial modeling. *Transactions of the Institute of British Geographers, 2*, 459–472.

Openshaw, S., & Rao, L. (1995). Algorithms for reengineering 1991 census geography. *Environment and Planning A, 27*, 425–446.

Parenteau, M. P., & Sawada, M. C. (2011). The modifiable areal unit problem (MAUP) in the relationship between exposure to NO2 and respiratory health. *International Journal of Health Geographics, 10*(1), 1–15.

Polani, D. (2002). Measures for the organization of self-organizing maps. In *Self-organizing neural networks* (pp. 13–44). Heidelberg: Physica-Verlag.

Shannon, C. E. (2001). A mathematical theory of communication. *ACM SIGMOBILE Mobile Computing and Communications Review, 5*(1), 3–55.

Vaz, E. (2018). Regional intelligence: A new kind of GIScience. *Habitat International, 72*, 1–108. https://doi.org/10.1016/j.habitatint.2017.11.015

Vaz, E., Cusimano, M., & Hernandez, T. (2015). Land use perception of self-reported health: Exploratory analysis of anthropogenic land use phenotypes. *Land Use Policy, 46*, 232–240.

Part IV
Resilience and Innovation

Chapter 10
Innovative Urban Paradigms for Sustainability and Resilience

Manuela Pires Rosa

10.1 Introduction

Nowadays the anthropic activities are disturbing the services that the natural ecosystems offer to society in terms of flows of materials, energy and information. Considering the scale of the global ecosystem, it is thought that we have come to a situation in which the capacity of regeneration and assimilation is at stake (Goodland and Daly 1996).

The models of spatial and social organization implemented, the population growth and the rise of the utilization of natural resources, have transformed the earth's surface, altered the biogeochemical cycles and modified the biological condition of the ecosystems, resulting in environmental and ecological problems, such as climate changes and the loss of biological diversity (Vitousek et al. 1997). These are associated with relevant social and economic impacts.

With this chapter, we advocate that sustainability and resilience perspectives are demanding new models of spatial and social organization, to which urbanists must pay careful attention. Deep changes are demanded in the management of the territory with particular attention to innovating approaches that stress the integrated structure of the social-ecological system, by promoting green and blue infrastructures and their ecosystem services.

We present the conceptual evolution of the sustainability and resilience perspectives, then we establish the implications to the urban management in a substantive and a procedural way and finally, we conclude with some of the best practices that are occurring in European cities.

M. P. Rosa (✉)
Research Centre for Spatial and Organizational Dynamics—CIEO, University of Algarve, Faro, Portugal
e-mail: mmrosa@ualg.pt

H. Pinto et al. (eds.), *Resilience and Regional Dynamics*, Advances in Spatial Science, https://doi.org/10.1007/978-3-319-95135-5_10

10.2 Sustainability

Sustainable development perspective was presented by the World Commission on Environment and Development, in 1987 (WCED 1987). It appears to finder a harmony between the economic development and the conservation of natural resources. The term "sustainability" traduces the quality of maintenance of something, which can continue for indefinite time, like, for example, biological species. It expresses a skill connected with the dynamic balance and with the interdependence between the natural ecosystems and the human ones, leading to its temporal maintenance. Consequently, it incorporates a multiplicity of dimensions, all of them inter-related: ecological, environmental, social and cultural, economic, territorial, institutional, political, governmental and individual (Rosa 2013, 2014).

The ecological dimension of sustainability has as aim the maintenance of the natural capital which is represented by the stock of natural resources, atmosphere and hydrosphere, ecosystems and species. The ecosystems should be kept healthy, by maintaining its "ecological integrity", i.e., the capacity of preserving the structure and functions of the communities. For this purpose, it is important to maintain the ecological diversity, resistance and flexibility, as well as the systems and functions of support of life, and the maintenance systems of the biochemical cycles. The ecological integrity is associated with the capacity of "resilience" of the ecosystems, which portray their aptitude to reorganize themselves when faced with a serious perturbation and continue their processes of self-organization (Rosa 2014).

Environmental dimension aims at managing the natural resources, so as to satisfy three basic generic lines proposed by Herman Daly (1990):

- The rates of use of renewable resources do not exceed their rates of regeneration;
- The rates of use of non-renewable resources do not exceed the rate at which substitutes are developed;
- The rates of pollution do not exceed the assimilative capacity of the environment.

Some related goals have to be considered: to conserve and to improve the basis of the natural resources; to conserve the landscapes stability; to conserve and improve the hydric resources; to protect the atmosphere in the regional and world scales; to conserve and improve the quality of the local environment (Rosa 2013). The need of reducing the dependence of non-renewable resources is stressed, as is the case of fossil fuels. In this process, it seems to be indispensable to have a climate stability (Goodland and Daly 1996). In this environmental context, it is fundamental to be aware of the ecosystem services, which are subdivided into four groups: support services, production services, regulation services and cultural services. These services result in benefits that people and organisations receive from the ecosystems and constitute well-being determinants (Millennium Ecosystem Assessment 2005). For example, including basic materials for a good life (health, good social relationships, security, freedom of choice and action).

The economic sustainability has as a goal the potential of production of the societies, which includes the stock production, distribution and transactions of the market (Rosa 2014). The production systems must preserve and increase natural capital by controlling finite stocks and balancing renewable resource flows. They should improve with technologies and processes, which lead to a more efficient, and less utilization of the natural resources and that produce fewer residues. This circular economy aims to keep products, components, and materials at their highest utility and value at all times (Ellen Macarthur Foundation 2015). To face the basic needs of all citizens, one should eradicate global poverty, by incentivising a "reasonable" rise of the economic growth (especially in the developing regions) and provide a more balanced distribution of costs and benefits. One must follow principles of solidarity in the distribution of natural wealth.

Social sustainability is mainly related with the concept of "equity", understood on a broader way, which goes beyond the equitative distribution of richness (Serageldin 1993). This is associated to the equality of opportunities of access to goods and services, to the access and management of information, to the development of local capacities, to the shared leadership and to the participation of different groups in decision-making, in a governance process (Rosa 2014). Societies must be informed, participative, and capable of awakening a sustainable development, according to its technologies, values, cultures and aspirations. Social sustainability is based upon the domain of the values and upon the cultural identity, which shape the mentalities and local conceptions that reflect life styles. With the cultural sustainability we aim at respecting the different cultures and their contributions for the construction of models of development appropriate to the characteristics of the communities and ecosystems, which integrate them. It is based on the respect for the endogenous roots of the communities, the affirmation of the local scale, together with the globalization (Rosa 2013).

Territorial dimension of sustainability has as aim territorial cohesion, which looks for a greater equity in the inter-regional relationships. We need new territorial models that detain a more balanced rural-urban configuration and a better territorial distribution of human settlements and economic activities (Sachs 1993).

The institutional, political and governmental sustainability has as goal the necessary organizational potential for knowledge, performance, administration and management of the government and administrations, being essential for the effective management of the social-ecological system (Rosa 2014). It is linked with the safety of all citizens and considers the risk society. In all this sharing process of responsibilities, one also requires profound changes in the pattern of consumption of the citizens, as well as their participation in the life of the communities and in decision-making. The individual sustainability describes the potential of the actions which enable and lead to attitudes and practices according to sustainability. The involvement of the citizens, in an active way, in the development process, is the key to success.

Sustainability is the relation between the dynamic human economic systems, and the ecological, also dynamic system, but which normally changes at a slower pace, in which: (a) human life can continue indefinitely; (b) human individualities can prosper; (c) human cultures can develop; but in which: (d) the effects of human activities stay within certain limits, so as not to destroy the diversity, complexity and functions of the ecological system which serve as support to life (Costanza 1997). Resulting from its rising importance, at the beginning of the twenty-first century, it emerged as a new scientific area, constituting the Science of Sustainability.

10.3 Resilience

The awareness of the character, mainly anthropocentric of the sustainable development, and the recognized co-evolution which exist between the social and the ecological systems, took some academics to defend that the perspective of "sustainability" was overtaken, in the nineties of the twentieth century, by the perspective of "resilience" (Leeuw and Aschan-Leygonie 2000) which determines the importance of change as a means of survival before situations of adversity and which claims for the attention of complexity and diversity of the socio-ecological system. Resilience is increasingly considered as a perspective or as a way of thinking to analyze linked social-ecological systems and to address social processes, such as social learning, leadership and adaptive governance (Folke 2006).

Decades ago, the concept of "resilience" was used in ecology by Holling (1973) who defines it as a measure of the persistence of systems and of their ability to absorb change and disturbance and still maintain the same relationships between populations or state variables. The ecosystems can reorganize themselves, i.e. they can move from one stable domain to another. This ecological dynamics, as an attribute, is translated by resilience and is inherent to a capacity to learn, self-organize, and evolve with change. According to Walker et al. (2004) resilience is the capacity of a system to absorb disturbance and reorganize it while undergoing change, so as, to, even so, retain, essentially, the same function, structure, identity, and feedbacks. In ecological science, resilience comprises three characteristics: (1) the capacity to absorb disturbances—the amount of change a system can undergo and still remain within the same domain of attraction, i.e., to retain the same controls on structure and processes; (2) the capacity for self-organization—the degree to which the system is capable of self-organization; and (3) the capacity for learning and adaptation—the degree to which the system expresses capacity for learning and adaptation (Folke 2006). This interpretation of resilience focuses on the robustness of systems to withstand shocks while maintaining function.

Resilience is not an exclusive feature of natural ecosystems, it has also been applied to artificial systems, whether economic, social or social-ecological systems. Social-ecological resilience is the capacity of a social-ecological system to absorb recurrent disturbances so as to retain essential structures, processes and feedbacks (Adger et al. 2005). This refers to the interplay of disturbance and reorganization within a system as well as to transformability, learning and innovation. Social-ecological resilience demands societies to adopt a systematic learning approach that attends to the way they interact and manage the natural environment (Olsson et al. 2014). Some of the concepts drawn from ecological resilience are being used to examine and manage social-ecological systems, such as resilience, adaptability, transformability, flexibility, diversity, robustness, redundancy, decentralization, etc. These attributes will influence their abilities to adapt to and benefit from change in a stable dynamic process. Adaptability is the capacity of actors from a system to influence resilience. In the case of social-ecological systems, this amounts to the collective capacity of the human, so it is a social function, where the individuals and groups act to manage the system (Walker et al. 2004). It expresses the ability to find new equilibrium levels to long-term changes in the environment. The adaptive capacity (or flexibility) is an aspect of the resilience that reflects innovation, learning and the ability to experience and adopt multiple and innovative solutions and develop generalized responses to a wide range of changes. Transformability is the capacity to create a fundamentally new system when ecological, economic, or social structures conditions make the existing system untenable (Walker et al. 2004).

Some academics recognized that diversity is linked to the stability of ecosystems. A decrease in diversity reduces the possibility of the system to cope with unforeseen circumstances, therefore, it is considered that diversity is very important in maintaining ecosystem resilience (Holling 1973). Robustness is the strength or the ability of systems or elements to withstand a given level of stress or demand without suffering degradation or loss of function (Bruneau et al. 2003) i.e., preservation of particular characteristics occurs despite uncertainty in components or in the environment. So a system is robust when it continues functioning in the presence of challenges without fundamental changes to the original system. Redundancy is associated with the extent to which systems or elements exist that are substitutable, so they are capable of satisfying functional requirements in the event of disruption, degradation or loss of functionality (Bruneau et al. 2003). It is an attribute associated to decentralized systems characterized by the reduction of exposure and vulnerability during disasters, providing robustness.

Conceptual analysis of sustainability and resilience perspectives permit to understand that they are intrinsically connected, one influencing the other (Fig. 10.1). Thus, perspectives of "sustainability" and "resilience" are complementary, it is not a question of replacement. Sustainability emphasizes a human development that attends to the conservation of natural resources and to the maintenance of ecological integrity which depends on the ecological resilience. It emphasises ecological and cultural diversity and promotes public participation and learning. The resilience perspective adds concepts related with the complexity of the social-ecological systems and deals with the present risk society. It gives significance of the evolving process that society needs to deal with a changing world that must be reorganized and adapted. This conceptual perception can be useful to the development of public policies or management processes of integrated urban systems, which as social-ecological systems are characterized by a nonlinear behaviour and increasing uncertainty.

Some academics have argued that the resilient paradigm took the place of the sustainability one (Leeuw and Aschan-Leygonie 2000; Cascio 2009). Others consider that both are interdependently linked. Resilience is related to both risk and sustainability, considering future generations (Blake 2013). After a disaster, the long term recovery period is a challenging time for resilience. In this period, resilience

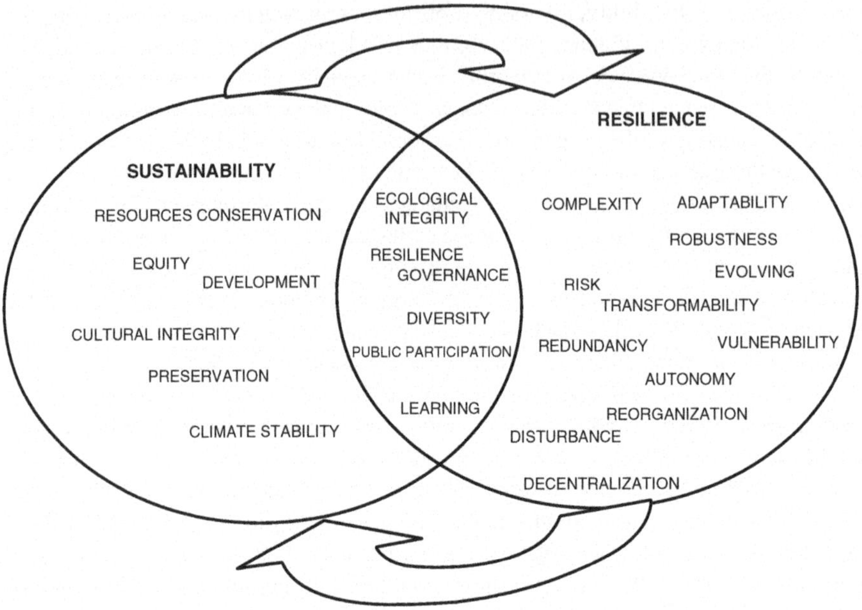

Fig. 10.1 Conceptual basis of sustainability and resilience perspectives. Source: Own elaboration

and sustainability become intertwined, as society becomes more resilient (e.g. more adaptable to future adverse events) and also sustainable (e.g. ensuring that future generations can survive and thrive) over the long term (Saunders and Becker 2015).

10.4 Towards Urban Paradigms Changes

In this context of sustainability and resilience, social movements and the academic world have been proclaiming the need of a New Economy, a New Urbanism and a New Mobility.

The movement for a New Economy appeared in England in the eighties of the twentieth century, defending an economy based on stability, sustainability and equality. It counts on, presently, with a broad network of organisations, for example, New Economics Foundation, New Economics Institute, New Economy Network, Tellus Institute, New Economy Working Group. These organisations are formulating a new economic theory aimed at the collective well-being, which considers the environmental preservation, the limits of natural resources, the social inequalities and the human well-being (Stephen et al. 2010). They value a territorial approach which enhances the endogenous resources: physical, environmental, cultural, human, economic and financial, institutional and administrative. They assume that such a territorial approach requires a complex, systemic, integrated and global, social change.

Expansionist theories associated with economic development strongly influenced land use planning, water and transportation policies through the implementation of large-scale engineering projects. These public policies, in force during the twentieth century, paid attention to the approach "predict and provide" (Owens 1995) the territory with infrastructures, based on measures of supply flexibility which betted on a continuous urban expansion and on water, energy and transport supply systems, according to the studied predictions. In the context of sustainability one requires changes of substantive tenor, which consider, in a complementary way, demand management measures which aim at managing (reduce or reorganize) certain needs (water, energy, new urban space), instead of just satisfying them.

In the planning and management of the water resources, one has paid attention to the "hydraulic paradigm" (Moral and Saurí 1999), which has stressed measures of supply management, through the construction of big dams and water systems. Presently one advocates a "New Culture of the Water" (Jiménez and Martínez-Gil 2005) which defends that rivers, lagoons and damp areas cannot simply be considered as mere ducts or deposits of water at the service of the productive systems, but they should be managed as live ecosystems and as collective patrimonies of identity, of natural well-being, beauty and evocation of feelings. It assumes an ecosystem approach and considers the relationships of water with human spiritual facet, therefore proposing a certain water ethics. The necessary ecosystem approach stresses the need of integration of water management with territorial management and it promotes adequate land uses (contributing to the reduction of human water needs) and the rehabilitation of the ecosystem services. These take advantage of the obtained benefits of the regulation of the processes of the ecosystems, as for example, climate regulation, floods control, aquifers recharge and the maintenance of water quality.

In practice, until a fairly recent past, in some European countries, territorial planning hasn't been able to configure itself as a preventive instrument of environmental management, but as an instrument of urban development, strongly influenced by expansionist theories. With this instrument the best localizations for the urban expansion, for industry, for leisure areas were found and transportation networks were implemented to connect the cities, without considering disaster risks. In this process, the territorial managers detained a building vision of the territory. These territorial models are a legacy of an abundant fossil fuel period which allowed a significant tendency to low demographic densities and a higher physical separation and dispersion of activities (work, residence, commerce, education and leisure), leading to a broad motorized mobility, which consumed a lot of energy and generated greenhouse emissions, among other impacts.

A New Mobility is required (Rosa 2013), i. e., a sustainable mobility, which must contribute to the territorial cohesion and social equity, not globally dependent on non-renewable natural resources and not putting public and ecosystems health in danger (adapted from OECD 1997). For this purpose we need to reduce the current motorized traffic (Broaddus et al. 2009) for a real decrease of the consumption of non-renewable resources and avoid the emission of greenhouse gas and other polluters. All this requires following the approach denominated "predict and prevent" (Owens 1995) in which one predicts a future search for dislocations on road and then one finds ways of avoiding that to happen, through new approaches of "getting the price right" and influencing travel patterns through land use planning. This transport demands management approach, it applies strategies and policies to reduce travel demand (specifically that of single occupancy of private vehicles), or to redistribute this demand in space or in time (Nelson 2000). For this purpose, the measures applied are: land use and transport development, public transport integration, parking controls and management, regulatory controls, physical measures such as bus and pedestrian priority, pricing and charges through fuels, annual taxes and congestion charging (Broaddus et al. 2009). Presently in many urban areas (Fig. 10.2) the environment for the pedestrian and cyclist is extremely hostile, due to the urban design itself, orientated for the motorcar traffic.

The ideal is to promote a balance of all relevant transport modes (Rupprecht Consult 2014), and the creation of a network of pedestrian and bicycle routes, accessible to all, associated to a green structure, inviting the citizen to a dislocation on foot or bike (Fig. 10.3). Concerning sustainable mobility, while traditional

Fig. 10.2 Inadequate pedestrian infrastructures. Source: Photos by the author

Fig. 10.3 Friendly urban environment for pedestrian and cyclist. Source: Photos by the author

transport planning approaches focuses on the movement of cars (particular transport modes) by expanding road infrastructures, now the emphasis should be laid on mobility and accessibility for all population groups. There is a focus on people (Rupprecht Consult 2014). These new approaches promote transit oriented developments that assure the access between residential areas and the stops and stations of public transport. These must be attractive for the pedestrian and for the cyclist, in terms of aesthetics, comfort and safety.

Considering the district scale, the pedestrian is at the top of the access hierarchy, including people with reduced mobility (Fig. 10.4) so that the living quarters become more human places.

The New Mobility is intrinsically connected with the New Urbanism. This is an urban design movement focused on the human scale, urban aesthetic environment, building patrimony and quality of life. All these attributes contribute to the creation of sustainable and competitive cities. It promotes mixed urban use (improving diversity), by encouraging the proximity of the urban services, work places and residential areas and, thus, creating short-distance cities. It is important that most residents live within 300 m of a green area or a stop of public transports. The ideal residential densities for sustainability are 300 inhab/ha (Fulford 1996).

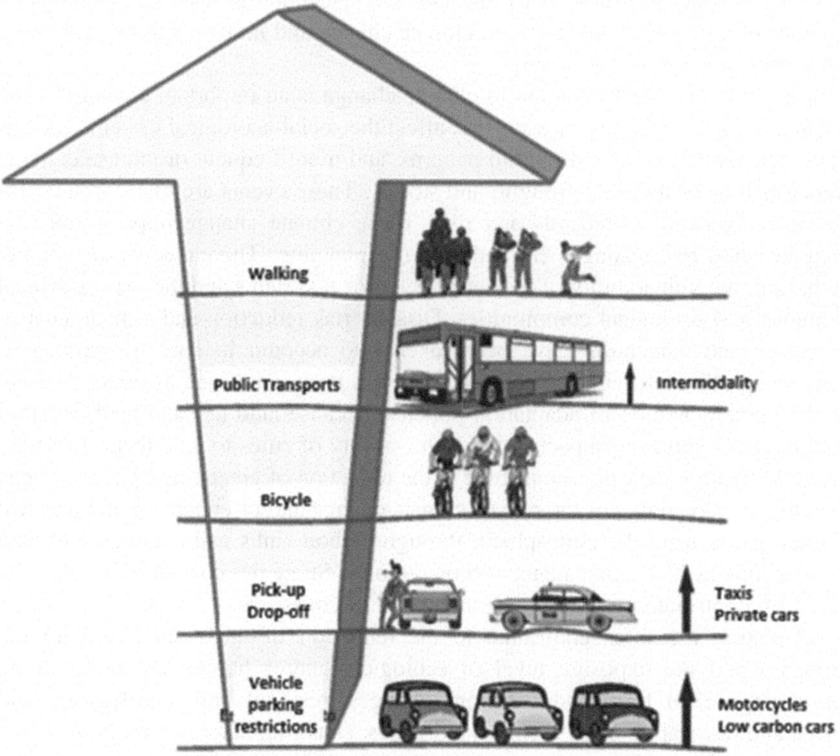

Fig. 10.4 Urban access hierarchy. Source: Own elaboration

A study from the International Association of Public Transport indicates an ideal density of 100 inhabitants and employers per hectare to promote walking, cycling and the use of public transports (Vivier et al. 2005). This sustainable urbanism promotes a new energetic paradigm by encouraging energy savings (reduce demanding), by maximizing the use of renewable energy sources and/or giving more attention to bio climate architecture and using fossil fuels in the cleanest possible way. Such changes contribute to low carbon and healthy cities. The goal is to create urban circular metabolism that minimize new energy inputs and maximize recycling. In this process, digital technology, associated with smart districts and cities, enables the reduction of energy consumption through intelligent transportation systems, public lighting and smart grids. Smart grids can improve peak load reduction, load shifting, co-generation/storage, the optimization of energy consumption, the reduction of costs through energy and operational efficiency, the reduction of greenhouse gases, the integration of renewable energy and promote sustainable e-mobility charging. This efficacy is related with smart and sustainable buildings that manage optimally local consumption, generation and storage, by providing detailed monitoring. The New Urbanism encourages the filling of the urban perimeters, imposing limits on urban growth, with processes of urban regeneration, which attend to energetic and water efficiency strategies, urban ecological improvement, and it also considers, in a preventive way, the natural and technological risks. The scientific evidence of a probable link between climate change and human activity provides a major challenge to policy-making.

In this domain, the adaptation to climate change is an important challenge. The global climate is changing in ways that affect the social-ecological systems such as higher temperatures, altered rainfall patterns, and more frequent or intense extreme events such as heatwaves, drought, and storms. These events are likely to increase the city's exposure to hazards and risk. These climate change impacts must be integrated into risk management and land use planning. The elements of risk are: the hazard, the vulnerability of the social-ecological systems and the exposure level of human and ecological communities. Disaster risk reduction and climate change mitigation and adaptation have been taken into account in land use strategies. Concerning climate change, spatial planning has been conceived as being decisive for the implementation of adaptation policies, because land use and land development have a significant impact on the vulnerability of cities to its effects (Bulkeley 2013). Mitigation measures contribute to the reduction of greenhouse gas emissions allowing the slowdown of the rate of climate change and/or enhancing the removal of these gases from the atmosphere, through carbon sinks and/or green and blue infrastructures which can enhance carbon removal. Strategies considered in the New Urbanism contribute for an effective reduction of emissions.

Adaptation measures contribute to the reduction of the vulnerability of the territories and the exposure level of ecological and/or human communities by considering urban form and functions changes, resilient built construction and design and re-locating away from hazard zones, changing land uses in these.

Paton et al. (2013) suggest that planning (including land use and emergency planning) is an integral part of creating a resilient society. Identify and minimizing the risks posed by the building, its equipment and fittings, and the natural hazards of the area have to be integrated into disaster prevention strategies and into a preventive urbanism. Urban planning has an important role in the changing of land uses of threatened areas, by improving and increasing green and blue infrastructures, because of their huge potential for the reduction of various risks. As urban ecosystems, these infrastructures also contribute to the reduction of the greenhouse gas emissions. They provide important regulating ecosystem services such as climate regulation (source of and sink for greenhouse gases, influence local and regional temperature, precipitation, and other climatic processes), hydrological flows regulation (groundwater recharge and discharge), water retention, water purification and waste treatment (removal of excess nutrients and other pollutants), erosion regulation (retention of soils and sediments) and natural hazard regulation (flood control and storm protection) (adapted from Millennium Ecosystem Assessment 2005). Natural wetlands contribute to the robustness of the cities, considering that they are part of an integrated approach to coastal and fluvial flooding which must be attended in land use planning. They can slow the runoff of water and increase the sub-surface and underground flow and mitigate the vulnerability of coastal areas with the rising sea level.

A resilient city values ecosystem services and needs to enhance diversity and ecological functions. It requires the development of environmental policies and participated management processes to protect and restore urban ecological processes that support ecosystem services. Other green measures have been implemented to reduce flood risks: from sponge roads to rooftop gardens, cities are investing in ways to harvest rainwater. On these permeable cities, almost every raindrop is captured, controlled and reused. In a climate change adaptation context, these "soft engineering" measures (Rosa 2013) will promote long-term robustness and more flexibility to adapt in the future. The application of the ecosystem approach in cities management will help to achieve a balance between conservation, sustainable use of resources and equitable distribution of benefits. This approach demands the collaboration of the citizens, it relates people to ecosystems and it needs an organizational and leading culture (governance), it achieves a social structure which takes horizontal and inverted decisions, aiming at an adaptive organization based upon a learning process.

10.5 European Sustainable and Resilient Urban Best Practices

Europe is nowadays an essentially urban society, with more than two thirds of European citizens living in towns and cities. Sustainability is a significant perspective shared by the European Union, member states and local authorities and it is taken into account in public policies. Promoting integrated sustainable urban development has been considered a key element of the European policies and a continuous process.

Sustainable development has long been on the political agenda of the European Union (EU), with the creation of its Sustainable Development Strategy. This strategy considers the economic, environmental, social, institutional and global dimensions. It promotes the adoption of good governance practices in the EU and the promotion of a global partnership for worldwide sustainable development. Ten thematic areas have been developed: (a) socioeconomic development; (b) sustainable consumption and production; (c) social inclusion; (d) demographic changes; (e) public health; (f) climate change and energy; (g) sustainable transport; (h) natural resources; (i) global partnership and (j) good governance.

The present Europe 2020 Strategy, adopted by the European Council on 17 June 2010, aimed at creating a smarter, greener and more inclusive economy and society. Climate change and energy constitute one of the headline targets considered in this strategy: reducing greenhouse gas emissions by at least 20% compared to 1990 levels; increasing the share of renewable energy in final energy consumption to 20%; moving towards a 20% increase in energy efficiency (European Commission 2010a).

In 2005, the European Commission mentioned the need for adaptation, in the Communication 'Winning the Battle against Global Climate Change', while it was encouraging Member States to take adaptation policies. A strategy on adaptation appears on the European agenda in 2007 (European Commission 2007) and, after 2 years of a discussing process, the White Paper on adapting to climate change was presented (European Commission 2009). It pointed out the cross-border dimensions of climate change impacts and adaptation measures. This was the basis for the adoption of the EU's Strategy on Adaptation to Climate Change on 16 April, 2013, in line with the European Strategy 2020. This Adaptation Strategy is intended to act as a comprehensive framework that will help the EU to make the transition to a low-carbon, climate-resilient economy. It stresses the need to improve the capacity to respond to the impacts of climate change at all levels of political power (EU, national, regional and local) through a coherent and coordinated approach (European Commission 2013). It demands climate resilience amongst the most vulnerable sectors to be increased.

Uncertainty surrounding trends in greenhouse gas emissions and the unpredictable nature of climate change impacts nourish some Member States inertia (Heras 2015). However, some local authorities (cities and towns) have prepared adaptive strategies because there is the perception of the effects of global warming in the territories. Many European cities have begun developing adaptation strategies or action plans. So, urban resilience perspective is creating a lot of strategies and good practices in development in the European Union.

In 2013, the European Commission Directorate General Climate Action published a report about Adaptation Strategies for European Cities, which analyzed the current adaptation strategies and measures developed in several European cities (Ricardo-AEA 2013). This project applied the Performance Acceleration through Capacity-building Tool which is one tool for assessing the capacities of organisations to address climate change adaptation. Its elements are nine organizational capacities necessary for adaptation: awareness, agency, leadership, agents of change, working together, learning, managing operations, programme scope and coherence, and expertise and evidence. This EU Cities Adapt project specifies the adaptation actions most often in cities' strategies:

- Individual construction measures (e.g. flood barriers, improvement of the drainage system);
- Promoting research projects to improve city staff's knowledge;
- Specific risk management or heatwave plans or changes in certain (planning) standards;
- Increasing public communication;
- Strategic design and use of green and blue infrastructure (green spaces and water bodies).

Thus, these measures focus on strengthening research and increased knowledge, communication and awareness of the population and urban planning and management for heat stress, drought, marine and/or river flood and storm water run off (Table 10.1).

It gave emphasis to the necessity of the land use planning to reduce floods risks, to promote green roofs and walls, public green areas and urban farming and gardening, to increase the capacity of water retention and storage, to consider raise albedo and the reduction of hardened surfaces and to provide shading. It seems that a valuing of "soft engineering" occurs and tends to complement "hard engineering" in terms of importance (Rosa 2013).

There is a great receptivity among local authorities to implement sustainable and resilient initiatives in EU. Some of the cities have been recognized for their good practices by awards. One of the policy tools the European Commission is using to address all these related challenges is the European Green Capital Award, which recognizes and rewards local efforts to improve the environment, the economy and the quality of life in cities. The cities are evaluated by an international expert panel through a detailed technical assessment of 12 indicators covering: ambient air quality; climate change, mitigation and adaptation; eco-innovation and sustainable employment; energy performance; green urban areas incorporating sustainable land use; integrated environmental management; local transport; nature and biodiversity; quality of the acoustic environment; waste production and management; wastewater treatment; and water management.

Nine cities have been awarded with the title of European Green Capital since its origin in 2010. Stockholm (Sweden) won the inaugural title, followed by Hamburg (Germany) in 2011, Vitoria-Gasteiz (Spain) in 2012, Nantes (France) in 2013, Copenhagen (Denmark) in 2014, Bristol (United Kingdom) in 2015, Ljubljana (Slovenia) in 2016, Essen (Germany) in 2017 and Nijmegen (The Netherlands) is holding this title for 2018.

Considering that we assume the complementary of sustainable and resilience paradigms, we are going to present the cases of Stockholm and Hamburg. These cities have a tradition in higher ecological conservation patterns displaying continuous good practices in sustainable urban development, paying attention to what their citizens and enterprises want, valuing their participation and innovative solutions in what concerns environmental challenges, and their integrated resilient measures.

Table 10.1 Summary of reviewed adaptations options

No.	Type of adaptation option	Heat stress	Drought	Flooding (marine)	Flooding (river)	Storm water run off
1	Construction and design of buildings	x				
2	Orientation of buildings and open spaces	x				
3	Green roofs and walls	x				x
4	Raise albedo	x				
5	Provide shading	x				
6a	Reinforce flood protection infrastructure (river)				x	
6b	Reinforce flood protection infrastructure (sea)			x		
7	Flood proof infrastructure			x	x	x
8a	Innovative flood protection options (river)				x	
8b	Innovative flood protection options (sea)			x		
9	Enhancing capacity of water storage				x	x
10	Geothermal heating and cooling	x				
11	Public green areas	x				x
12	Urban farming and gardening	x	x			x
13	Land use planning to reduce floods risks			x	x	
14	Flood forecasting and warning systems			x	x	x
15	Heat health warning system	x				
16	Improve regulations for building			x	x	x
17	Evacuation and contingency management plans	x		x	x	x
18	Water saving measures		x			
19	Crisis management		x	x	x	x
20	Extend water supply services	x				
21	Floating and amphibian housing				x	
22	Public education and awareness campaigns	x	x	x	x	x
23	Reduce hardened surfaces					x
24	Compartmentalization			x		
25	Water management plans		x	x	x	x
26	Water retention		x		x	x
	Total	12	6	11	13	13

Source: Ricardo-AEA (2013, 32 and 33) Reprinted with permission of European Commission Directorate-General for Climate Action

The City of Stockholm has a large and varied ecological structure, around 40% of the city's land consists of parks and green spaces. More than 90% of the population lives within 300 m of a green area. Stockholm has very low greenhouse gas emissions because of its energy efficiency strategies: high proportion of renewable energy for heating houses, implementation of fibre optics, reduction of motorized traffic, cleaner vehicles and green electricity, involvement of clean-tech companies. Their targets is to become fossil-fuel free by 2050. It has an extensive and well developed public transport system, 90% of the residents live within 300 m of this, and during peak hours, 78% of all trips to the inner city are made by these means. Within the city center, 68% of all trips are made on foot or by bicycle (European Commission 2010a). These sustainable measures contribute for the mitigation of climate change (fewer greenhouse gas emissions) and to the capture of carbon (one of the ecosystem services from green and blue infrastructures). The city developed a climate change adaptation policy, in 2005, and developed an action programme, incorporating a study on adaptation to climate change, to identify the impacts of climate change in the city and to provide a foundation to adapt to these impacts (Ekelund 2007). Considering the potential role of spatial planning for climate change adaption through adaptation strategies (avoidance and minimization), studies conclude that Stockholm is an example of the implementation of these ways of adaptation in the strategic and detailed planning stages (Davidse et al. 2015). According to these authors, four major issues have been considered: (1) rising sea levels (the land is elevated where necessary); (2) heavy rainfall (green roofs, porous surfaces, green spaces); (3) heat and drought (green spaces network guaranteed shadow, water ponds and wetlands for irrigation) and (4) ecosystem quality (contributes for biotic processes to avoid urban heat island effect and to create a good micro climate).

In Hamburg, parks and other green spaces cover around 40% of the city's surface corresponding to 38 m^2/inhab and 89% of the residents live within 300 m of a green area. Urban renewal in the inner city (ex. the HafenCity) attended to the promotion of sustainable buildings and enabled residents to live and work in the city center, thereby reducing commuting levels and environmental problems caused by traffic or urban sprawl. There is a strong governance that highlights the special relationship between the environment and industry. In order to protect the climate, the city aims to reduce carbon dioxide emissions by 40% by 2020, and by 80% by 2050. All the residents live within 300 m of public transport (European Commission 2010b). As the case before, these sustainable measures contribute for the mitigation of climate change (fewer greenhouse gas emissions) and to the capture of carbon (one of the ecosystem services from green and blue infrastructures). In December 2015, the Senate of Hamburg approved its new climate plan, which also integrates adaptation for the first time. A strategy-based framework for assessing the flood resilience of the city was developed with the identification of components to implement resilience strategies. Aims to move from definition to "doing" resilience (Restemeyer et al. 2015). This framework shows that resilience requires capacity-building among public as well as private stakeholders. For example, Hamburg's green roof programme, from the municipality, supports building owners to establish green roofs. This measure will retain excess water and delay its entry into the rainwater drainage system and with lower costs (Ansel 2011). This measure was considered a good answer to climate change as it can improve the microclimate and enhance rainwater retention.

10.6 Final Considerations

Conceptual evolution analysis of sustainability and resilience perspectives permit a better understanding of the challenges we can meet for urban ecology and contribute to address social learning, governance and adaptive management. The associated concepts can be useful to the development of new public policies and management processes of integrated urban systems which are increasingly characterized by nonlinear behaviour and uncertainty.

Considering these attributes it seems that the contemporary models of spatial and social organization are no longer pertinent. Urbanists must adopt ecosystem approaches that take into consideration ecological and cultural integrity, and adaptive approaches which are associated with capacity building, social flexibility and learning, integrating the values and the perceptions of communities. They must undertake holistic territorial planning and management that aim at the integration of natural and social sciences and traditional conservation values.

In this context the need of a New Economy, a New Urbanism and a New Mobility has been proclaimed. The consideration of these changes is decisive to confront climate change and to avoid the increasing social inequality problems associated with the present risk society.

In practice, until a fairly recent past, in some European countries, land use planning hasn't been able to be itself as a preventive instrument of environmental and risk management. In this process, architect and civil engineers detained, dominantly, a building vision of the territory. Nowadays, the performance of the urbanists must be broader and holistic. Such should imply bringing green and blue infrastructures inside and around the cities, modifying urban form and function, and should demand a deep reformulation in energy, mobility and water systems, which must be decentralized for bigger urban robustness.

Cities such as Stockholm and Hamburg have all shifted perspective from risk to an opportunity to make the city more sustainable and resilient. The good governance of these paradigms demands effective local organisations and citizens and contributes to a territorial marketing. Learning from these examples, other cities may go through this progress faster.

Acknowledgments Author would like to express her thanks to European Commission Directorate-General for Climate Action and Praise Worthy Prize which granted permissions for the reprinting of information.

References

Adger, W., Hughes, T., Folke, C., Carpenter, S., & Rockström, J. (2005). Social-ecological resilience to coastal disasters. *Science, 309*, 1036–1039.

Ansel, W. (2011). *Kosten-Nutzen-Analyse Dachbegrünung, Deutsche Bundesstiftung Umwelt (DBU), Projekt-Nr. 28269–23*. Nürtingen: Abschlussbericht.

Blake, H. (2013). *The treasury's higher living standards framework: Update on the sustainability dimension*. Proceedings of the paper presented at the sustainability Rhetoric: Facts and fictions.

Broaddus, A., Litman, T., & Menon, G. (2009). *Transportation demand management: Training document*. Eschborn: GTZ.

Bruneau, M., Chang, S., Eguchi, R., Lee, G., O'Rourke, T., Reinhorn, A., et al. (2003). A framework to quantitatively assess and enhance the seismic resilience of communities. *Earthquake Spectra, 19*, 733–752.

Bulkeley, H. (2013). *Cities and climate change*. New York: Routledge.

Cascio, J. (2009, April). The next big thing: Resilience. *Foreign Policy*. Accessed July 10, 2016, from http://www.openthefuture.com/media/FPcascio.pdf

Costanza, R. (1997). La economía ecológica de la sostenibilidad. Invertir en capital natural. In R. Goodland, H. Daly, S. El Serafy, & B. Von Droste (Eds.), *Medio Ambiente y Desenrollo Sostenible – Más allá del Informe Brundtland*. Madrid: Editorial Trotta.

Daly, H. (1990). Commentary: Toward some operational principles of sustainable development. *Ecological Economics, 2*, 1–6.

Davidse, B. J., Othengrafen, M., & Deppisch, S. (2015, April). Spatial planning practices of adapting to climate change. *European Journal of Spatial Development, 57*. Accessed September 1, 2016, from http://www.nordregio.se/Global/EJSD/Refereed articles/refereed57.pdf

Ekelund, N. (2007). *Adapting to climate change in Stockholm*. Stockholm's action programme on climate change. Stockholm: Stockholm Stad (City of Stockholm).

Ellen Macarthur Foundation. (2015). *Towards the circular economy. Business rationale for an accelerated transition*. Accessed September 24, 2017, from https://www.ellenmacarthurfoundation.org/assets/downloads/TCE_Ellen-MacArthur-Foundation_9-Dec-2015.pdf

European Commission. (2007). *Green paper on adapting to climate change in Europe - options for EU action*. COM (2007) 354 final.

European Commission. (2009). *White paper: Adapting to climate change: Towards a European framework for action*. COM(2009) 147 final.

European Commission. (2010a). *Stockholm – European Green Capital 2010*. Luxembourg: Publications Office of the European Union, European Commission. ISBN: 978-92-79-15064-7.

European Commission. (2010b). *Hamburg – European Green Capital 2011*. Luxembourg: Publications Office of the European Union, European Commission. ISBN: 978-92-79-19009-4.

European Commission. (2013). *An EU strategy on adaptation to climate change*. Communication COM(2013) 216 final.

Folke, C. (2006). Resilience: The emergence of a perspective for social-ecological systems analysis. *Global Environmental Change, 16*(3), 253–267.

Fulford, C. (1996). The compact city and the market. In M. Jenks, E. Burton, & K. Williams (Eds.), *The compact city: A sustainable urban form?* (pp. 122–133). London: E&FN Spon Publishers.

Goodland, R., & Daly, H. (1996). Environmental sustainability: Universal and non-negotiable. *Ecological Applications, 6*, 1002–1017.

Heras, B. P. (2015). Adaptation to climate change in the European Union: Between supranational action and state reluctance. *Iucn Academy of Environmental Law EJournal, 6*, 1–21. Accessed September 12, 2016, from http://www.iucnael.org/es/revista-electronica-de-la-academia/numero-actual

Holling, C. S. (1973). Resilience and stability of ecological systems. *Annual Review of Ecology and Systematics, 4*(1), 1–23.

Jiménez, N., & Martínez-Gil, J. (2005). The new culture of water in Spain: A philosophy towards a sustainable development. *E-WAter Official Publication of the European Water Association, 7*, 1–20. Accessed in September 12, 2016, from http://www.ewaonline.eu/tl_files/_media/content/documents_pdf/Publications/E-WAter/documents/58_2005_07.pdf

Leeuw, S., & Aschan-Leygonie, C. (2000, May 22–26). *A long-term perspective on resilience in socio-natural systems*. Workshop System shocks – System resilience Abisko. Sweden. Accessed June 24, 2017, from https://sfiedu.s3.amazonaws.com/sfi-edu/production/uploads/sfi-com/dev/uploads/filer/4b/b6/4bb681ae-0b24-4866-8fb2-47e8b92e90c1/01-08-042.pdf

Millennium Ecosystem Assessment. (2005). *Ecosystems and human well-being: Synthesis*. Washington, DC: Island Press. ISBN: 1-59726-040-1.

Moral, L., & Saurí, D. (1999). Changing course: Water policy in Spain. *Environment, 41*(6), 12–36.

Nelson, D. C. (Ed.). (2000). *Intelligent transportation primer* (pp. 10–11). Washington, DC: Institute of Transportation Engineers. ISBN: 0-935403-45-0.

OECD. (1997). Towards sustainable transportation. In *Proceedings of the international conference*. Organized by the organisation for economic co-operation and development and hosted by the government of Canada, Vancouver, British Columbia, 24–27 March 1996, Environment Directorate. Paris: OCDE.

Olsson, P., Galaz, V., & Boonstra, W. J. (2014). Sustainability transformations: A resilience perspective. *Ecology and Society, 19*(4), 1. https://doi.org/10.5751/ES-06799-190401

Owens, S. (1995). From "predict and provide" to "predict and prevent"? Pricing and planning in transport policy. *Transport Policy, 2*(1), 43–49.

Paton, L. D., Mamula-Seadon, L., & Selway, K. (2013). *Community resilience in Christchurch: Adaptive responses and capacities during earthquake recovery*. GNS science report 37. lower Hutt: GNS science.

Restemeyer, B., Woltjer, J., & van den Brink, M. (2015). A strategy-based framework for assessing the flood resilience of cities: A Hamburg case study. *Planning Theory & Practice, 16*(1), 45–62. Accessed September 12, 2016, from https://doi.org/10.1080/14649357.2014.1000950

Ricardo-AEA. (2013). *Adaptation strategies for European cities*. Final report. ©European Union. Report prepared for the European Commission Directorate-General for Climate Action.

Rosa, M. P. (2013). Implications of complexity and sustainability in civil engineering. *International Review of Civil Engineering, 4*(1), 55–61.

Rosa, M. P. (2014). Integrated vision for sustainable and innovative industrial areas. In M. Cavallo, K. Konstantinou, & C. Makrakis-karachalios (Eds.), *Marketing theory and governing innovative industrial areas integrated guidelines – A handbook, MER project – marketing theory and governing innovative industrial areas (2013–2014)*. Accessed September 24, 2017, from http://www.informest.it/docs/post/WP3_Wiki_e-book.pdf. ISBN ebook 978-618-81809-0-1.

Rupprecht Consult. (2014). *Guidelines: Development and implementing a sustainable urban mobility plan*. Brussels: European Commission.

Sachs, I. (1993). *Estratégias de Transição para o Século XXI. Para Pensar o Desenvolvimento Sustentável*. São Paulo: Nobel. ISBN: 8585445112.

Saunders, W., & Becker, J. (2015). A discussion of resilience and sustainability: Land use planning recovery from the Canterbury earthquake sequence, New Zealand. *International Journal of Disaster Risk Reduction, 14*, 73–81. Accessed July 2, 2016, from http://www.sciencedirect.com/science/article/pii/S2212420915000230

Serageldin, I. (1993). Making development sustainable. *Finance and Development, 30*, 6–10.

Stephen, S., Simms, A., Neitzert, E., & Ryan-Collins, J. (2010). *The great transition*. London: New Economics Foundation. ISBN: 9781904882633.

Vitousek, P., Mooney, H., Lubchenco, J., & Melillo, J. (1997). Human domination of earth's ecosystems. *Science, 277*, 494–499.

Vivier, J., Pourbaix, J., & Mezghani, M. (2005). *Mobility in cities – Database*. Brussels: International Association of Public Transport.

Walker, B., Holling, C., Carpenter, S., & Kinzig, A. (2004). Resilience, adaptability and transformability in social–ecological systems. *Ecology and Society*, *9*(2), 5. Accessed September 2, 2017, from http://www.ecologyandsociety.org/vol9/iss2/art5

WCED. (1987). Our common future. In G. H. Brundtland (Ed.), *Report of the world commission on environment and development*. Oxford: Oxford University Press.

Chapter 11
Innovation as Transformation: Integrating the Socio-ecological Perspectives of Resilience and Sustainability

Karl Bruckmeier and Iva Pires

11.1 Introduction: The Social and Epistemic Contexts of Resilience and Sustainability

Innovation, the creation of new technologies in the spheres of knowledge use, business, policy or natural resource management, is a highly context-dependent and institutionally steered process. A paradox of innovation can be formulated as: innovation is both a cause for our current unsustainable trajectory and a hope for tipping in new more resilient and sustainable directions. Practically seen innovations are used to solve specific problems. These problems, especially environmental problems, are often of complex nature and require the integration of the technical innovation process itself with the political, economic or civil society action of many actors, institutions or social groups. Therefore, a technical innovation becomes, when it is applied, part of social innovations. The social processes of development and change show the main problems of innovations in natural resource management and environmental policies: the innovations require change or transformation of social behaviour of certain social groups and actors with different interests and aims. Because of the significance of the social components of innovation processes, we ask: What kind of behaviour changes and social transformation do environmental problems—that are today global, consequences of global environmental and social change—require? When innovation becomes part of overarching processes of problem solving and social change, it can be said, innovation becomes (part of) social

K. Bruckmeier
Department of Sociology, National Research University – Higher School of Economics, Moscow, Russian Federation
e-mail: kbrukm@hse.ru

I. Pires (✉)
Faculty of Social Sciences and Humanities, New University of Lisbon, Lisbon, Portugal
e-mail: im.pires@fcsh.unl.pt

H. Pinto et al. (eds.), *Resilience and Regional Dynamics*, Advances in Spatial Science, https://doi.org/10.1007/978-3-319-95135-5_11

transformation. In the case of environmental problems that have an impact on social and ecological systems simultaneously, the important context factors of innovation and change are resilience and sustainability, themselves complex processes of change.

The concepts of resilience and sustainability exist in different and incompatible versions in ecological research—as complementary or as contrasting concepts. The controversial discussion of the concepts requires their interpretation before they can be used in social or ecological research. Our interpretation of resilience and sustainability in the broader framework of coupled social-ecological systems (SES) is compatible with a widespread use, but competing interpretations exist. SES are theoretically conceptualised in social ecology (Fischer Kowalski and Haberl 2007; Bruckmeier 2013, 2016) as interconnected systems where ecological or social-ecological resilience means basically a capacity to adapt to disturbance (Folke 2006). Sustainability is, in contrast to resilience, seen as a more long-term process of transformation of SES that requires beyond adaptation to disturbance a capacity to initiate and maintain over long periods of time, decades or centuries a process of directed change. Such long periods cannot be planned and managed, but require further, more complex modes of steering and governance that allow to influence indirectly complex social and ecological processes that cannot be planned for—this is the real challenge of transformation towards sustainable resource use in modern society.

In the research on transition and transformation in recent years (for a summarising discussion see Markard et al. 2012), the criterion of sustainability is seen as that of changing the processes of exponential economic growth and growth of resource use, reducing them, ideally seen to zero growth. However, in the broad sustainability discourse after the Brundtland report from 1987, where sustainability was seen as intra- and inter-generational solidarity of resource use, no exact criteria have been agreed upon; the ideas of zero-growth or degrowth remain controversial up to now. Ideas of degrowth find support from ecological research (see the "limits to growth"—reports of the Club of Rome; Asara et al. 2015), but less from economic or other social research, where continuing growth is seen as compatible with environmental goals when simultaneously mechanisms to reduce pollution and degradation of ecosystems, are implemented, for example, in policies of ecological modernisation (Mol et al. 2009).

The integration of the two perspectives of resilience seen as adaptive process (to cope with disturbance) and sustainability as transformative process (to achieve a balanced interaction of social and ecological system components), aims at a broader, interdisciplinary and integrated perspective of innovation and adaptation than possible with the single concept of resilience. Clarifying the concepts of resilience and sustainability requires some further concepts that are connected to the use of both, especially that of risk and vulnerability.

11.2 Reframing Innovation: Connections Between Risks, Vulnerability, Resilience and Sustainability

1. Risk was since its origins in the economic and technical sciences a formal, probabilistic and calculable term in the sense of the probability of a negative event/consequence of action. This is specified in the classical term of risk in economics by Knight (1921) as outcomes of action for which insurance is possible, whereas uncertainty refers to outcomes of action for which no insurance possible. With the sociological risk research, especially by Beck (theory of risk society, 1986) and Luhmann (1991) in late twentieth century, risks are studied in a broader social context and as risks in new forms that cannot be formulated with the classical risk concept which refers to individual or organisational decisions. The concept of systemic risks which is described by Renn and Klinke (2004); it converges with Becks conceptualisation of risks that are non-calculable and require the analysis of the functioning and interactions of large-scale social and ecological systems. Some of these risks cannot be perceived (e.g., nuclear radiation), known only from science. Three variants of risk analysis can be differentiated in epistemological terms (Diekmann and Preisendörfer 2001: 58):

   – Realist variants: risk analysis of engineers where objectivity of risks is defined through probability of negative outcomes and quantity of damage;
   – Constructivist variants: cultural theory of risks of Douglas and Wildavsky: members of different cultures perceive/interpret risks differently, which implies that it is difficult or impossible to achieve a common understanding;
   – Variants in-between realist and constructivist, where risks include as well constructivist as objectivist components, culturally specific subjective and social perceptions of risks and dangers that exist objectively, thus understanding risks as real and as constructed phenomena. This seems adequate for many forms of environmental and systemic risks that are identified through science and research, but are perceived and interpreted differently by individuals or social groups, according to their knowledge, values, and interests.

2. Vulnerability can be understood as exposure to risks and dangers; risk and vulnerability seem concepts depending on each other. Vulnerability implies a broadening of risk analyses, as caused by social or ecological factors or disturbances. It means primarily social vulnerability. The vulnerable subjects are humans. Also when vulnerability is differentiated for biophysical and social systems (Füssel 2007), as caused by disturbances in natural or social systems, it remains in both cases social or human vulnerability, as highlighted in the review by Lundgren and Jonsson (2012). They discuss (referring to Cutter et al. 2003) social vulnerability through natural hazards or climate change, as dependent upon:

   – "Lack of access to resources (monetary, information, knowledge or technology)
   – Limited access to political power and representation

- Social capital (including social networks)
- Beliefs and customs
- Building stock and age
- Frail and physically limited individuals
- Type and density of infrastructure and lifelines".

Social variables important for vulnerability to climate change are seen as age, gender, race and socio-economic status (Lundgren and Jonsson 2012: 3).

The extension of the vulnerability concept to imply vulnerability of ecosystems or of the global social and ecological systems maintains the metaphorical connotation of vulnerability as a health related term; it remains a term that receives scientific and theoretical meaning more from that of risk, from which it develops. Resilience, in difference to vulnerability, is not referring to social actors, groups or communities; its theoretical meanings are derived from the functions of ecosystems or coupled social and ecological systems—more a system capacity than an action capacity.

3. For the further development of the resilience concept (Bruckmeier and Olsson 2014) two variants of resilience of ecosystems need to be discussed:
   - Resilience as maintaining or regaining stability after disturbance (engineering resilience), and
   - Resilience as capacity to absorb disturbance through shifts to other equilibrium states without collapsing (ecological resilience).

The second version of resilience can be broadened to include more complex types of coupled social and ecological systems (Folke et al. 2005; Folke 2006). With the broadening of the resilience concept to social system components, the meanings of resilience change further. Social and ecological forms of resilience are not parallel phenomena, but may imply contradicting requirements of maintaining social structures and functions or ecosystem functions (Adger 2000). Lloyd et al. (2013) specify social components of resilience in the notion of social-ecological resilience as capacity that implies social and transformative learning of social actors to support the restructuring of a SES in response to turbulence or catastrophes. In this sense the elastic concept of resilience that does not necessarily require the capacity of action and anticipation of humans, but can be based on simpler capacities of behaviour change, gains a clear social meaning. Social or collective learning is a core capacity for resilience and adaptation and for sustainability and transformation.

Connecting vulnerability and resilience analysis, with resilience as a capacity of social or ecological systems to cope with disturbance without collapsing, requires for coupled social and ecological systems (SES) the identification of strategies to enhance social-ecological resilience: These strategies can be:

- Strategies to reduce vulnerability through analysis of disturbance (identifying main disturbances for an area or social community), identifying crucial vulnerabilities through vulnerability assessment, mitigating vulnerability (through measures for reducing exposure to hazards and disturbance or compensating for their effects), reducing sensitivity (minimising responsiveness to changes through disturbance), institutional development (building and developing institutional capacity to prepare for disturbances and minimise their impacts), and trajectory management (oriented to projected changes relevant for future development: Chapin III et al. 2009);

– Strategies to enhance the adaptive capacity of the SES, e.g., fostering diversity, stabilising feedbacks and creative renewal, learning to live with change and uncertainty, adapting institutions and governance to changing conditions environmental conditions, building participatory and deliberative, developing multilevel governance through adaptive management or governance.

4. Sustainability is the most complicated and contested term used here. It implies in difference to resilience the maintenance of long-term development capacity of SES that cannot be reduced to the management of disturbance and crises. It implies, beyond resilience capacities, to cope with the limited availability of natural resources and redistribution of resources between users. Connections of different spatial and temporal scales are effective in the capacities of SES to achieve sustainability. In recent years has, after a long and often controversial debate of sustainability in science and policy, developed a new perspective that connects sustainability with the scientifically elaborated concept of socio-ecological transformation. This re-interpretation of sustainability (Bruckmeier and Olsson 2014) is used in the following analysis.

The challenges of transformation where the components of resilience and sustainability interact can be described in terms of three combined concepts and processes of *innovation* (Leach et al. 2012), *adaptatio*n (Armitage and Plummer 2010), and *transformation* (Raskin et al. 2010). These terms mark the complexity of processes of socio-ecological and socio-cultural transformation that cannot be reduced to political processes, although the "governance of sustainability" and its operational components such as "transformative action groups" are key components. Sustainability transformation or transition requires governance strategies for indirectly influencing the complex processes that work slowly and over long time, such as certain processes in ecosystems, or population growth and demographic transition. Incremental changes are not sufficient to cope with the prevailing challenges we face in several domains (energy production, water supply, pollution, greenhouse gas emissions, nuclear risks, extreme weather events); necessary are "long-term, multi-dimensional, and fundamental transformation processes" (Markard et al. 2012: 956). Fischer-Kowalski and Rotmans (2009) discuss different, micro- and macroscopic approaches to socio-ecological transition or transformation. Smith et al. (2005) describe four different strategies of transition that show the varying contexts of transition management in the governance of sustainability: endogenous renewal; re-orientation of trajectories; emergent transformation and purposive transitions.

Governance of sustainability requires, finally, a social-ecological theory of transformation that systematizes the analysis of spatio-temporal dynamics in coupled SES: an interdisciplinary theory that can be connected with empirical research and other theories, for example, theories of innovations. The dynamics of resilience and sustainability in SES can be summarised as:

- Adaption in a shorter temporal perspective where the reaction to disturbance and maintenance of balance and system boundaries after disturbance are the basic criteria (a dynamic derived from the functional mechanisms of ecosystems), and
- Transformation in a longer temporal perspective where the capacity of a whole society to change its systemic structures in coherence with the requirements of maintaining functioning ecosystems (a dynamic derived from the structures and processes of societal systems).

11.2.1 The Challenges: Connecting Analyses of Risk, Vulnerability, Resilience and Sustainability

Connecting risk, vulnerability, resilience and sustainability analyses is rarely done in one comprehensive system analysis of SES. For such an integrated analysis that can be done in several and separate parts, a series of decisions about the interpretation and application of the terms needs to be made, for which no exclusive support through scientific knowledge can be claimed. The notions discussed here—risk, vulnerability, resilience, sustainability—are elastic, have plural and competing meanings, and no consensus is available for their interpretation and application. The forms of such sustainability syntheses differ, but all of them have several common components. In difference to the widespread use of the resilience concept, for example by the "Resilience Alliance" and the Stockholm Resilience Centre, where also the interpretation of sustainability is dominated by the meaning of resilience as adaptation (adaptive cycles, as formulated in ecological research, Gunderson and Holling 2002), the conceptual and knowledge syntheses of the kind we discuss are based on the core concepts of transition or transformation.

The management of transition to sustainability is confronted with threefold transformation challenges:

1. Social challenges of rural-urban development in metropolitan areas: Processes of urbanisation happen today rapidly and are often badly managed, causing new social imbalances, inequity and poverty. The dimensions of megacities with many million inhabitants make cities less and less manageable. Cities are vulnerable through climate change and have increasing problems of food security. Mobility in form of migration or commuting to cities has become the dominant process in many countries. Increasing re-migration to the countryside indicates a crisis in urban development which requires new, integrated strategies of local development. Combined strategies of resilience and sustainability do not only require technical innovations and technologies, but social innovations that are created, for example, by transformation action groups and help to solve problems of social and environmental change.

2. Methodological challenges of "the city without boundaries": The administrative boundaries of cities are no longer effective for sustainable governance. Cities stretch through their resource use in the surrounding rural areas and far beyond—through the global flows, exchange and trade of resources they are becoming global in the sense that are dependent in their natural resource use from global flows of energy, matter and information. The global stretching of cities can be measured in the land and the resources they use beyond their boundaries. In local transformation strategies this global interdependence through resource flows needs to be reduced to create local opportunities for resilience and sustainability that support simultaneously the transformation at national and global levels. Resilience and sustainability require new ways to deal with the planetary boundaries of resource use at local levels, e.g., using methods and indicators as ecological footprint analyses of cities, material and energy flow accounting (MEFA), human appropriation of net primary production (HANPP) of ecosystems, energy return on input (EROI) in production and resource use processes.
3. Conceptual challenges—"the resilience and sustainability paradoxes": Resilience and sustainability are "essentially contested concepts" (Collier et al. 2006), defined and interpreted in many different ways. They became overused and over-interpreted, give no compass and guidance for sustainable transformations. The future sustainable society is unknown and has to be built with knowledge developed and experience gained in the transformation processes. At this point of navigating society into an unknown future innovation processes become "strategic variables": transformation implies social, cultural, technical and technological, economic, and political innovation. Transformation needs to be constructed anew, with measurable concepts for which the ideas of resilience and sustainability require re-interpretation with new knowledge and conditions of global change that affect local development. For cities, a paradox can be specified in the idea of the urban sustainability multiplier by Rees: cities do not have ecological disadvantages only, also a series of advantages that help to save material, energy, space and using resources effectively.

With this description of the nexus of resilience and sustainability, the interpretation of *resilience has the focus on adaptation (adaptive cycles)* and that of *sustainability implies the transformation of social-ecological systems* of modern society. Beyond adaptation to climate change and disturbances, sustainability requires a long-term, future-oriented perspective of development and collective action, resulting in changes in the interactions between social and ecological systems, the core processes determining sustainability. We illustrate combined forms of resilience and sustainability analyses with studies from rural and urban areas in European research projects.

11.2.2 Integration of Resilience and Sustainability Analyses in Studies of Rural–Urban Interaction

1. *Resilience related to agricultural production:* Agro-ecosystems and agricultural SES as our model cases are examples of interacting social and ecological systems. Beyond the basic meaning of resilience as adaptation to disturbance and system or boundary maintenance under conditions of stress, three types of resilience can be differentiated for these system types:

   – Ecosystem resilience advanced from studies of ecological resilience (Cabell and Oelofse 2012) to include ecosystem services or benefits provided by ecosystems to humans (Paavola and Hubacek 2013). Also alternative forms of agriculture, organic farming or community based agriculture, where the connection between ecosystems and people or resource users is emphasized (King 2008) can be understood as enhancing ecosystem resilience.
   – Livelihood resilience refers to people as part of SES. Rural livelihood studies have been carried out in great number and manifold perspectives, also regarding the connections between vulnerability, resilience and sustainability. Local livelihood studies

> "may miss out on long-term shifts which will, in time, undermine livelihoods in more fundamental ways. Long-term temperature rises may make agriculture impossible, shifts in terms of trade may undermine the competitiveness of local production or migration of labour to urban areas may eliminate certain livelihood options in the long-term. . . . Sustainability and resilience thus cannot always emerge through local adaptation in conditions of extreme vulnerability." (Scoones 2009: 19).

– Climate resilience refers to global climate change and its consequences for agriculture, especially in the Global South where the majority of agricultural producers are (poor) smallholders.

With these concretisations of the resilience terminology, resilience analyses can be developed as an interim step of sustainability analyses: resilience is one of the manifold processes to deal with in strategies of sustainability governance. Studies of rural-urban development and interaction—where it is necessary to connect a variety of social and economic development dynamics of different kind—show that resilience and sustainability require a systematic reconstruction of the system-maintaining processes in social and ecological systems which became especially complex with the continuing globalisations of economic and natural resource management processes.

2. *Rural-urban interaction in late modern societies—the consequences of globalisation:* In local strategies for rural development that is closely connected to metropolitan areas (for example, in the forms of peri-urban and urban agriculture, in metropolitan areas and their surroundings) we can study how the transformation processes unfold their dynamics in reaction to social and ecological change.

It is characteristic for such areas that a variety of specific development processes that unfolded their own dynamics in the course of history and modernisation, are interlinked, overlapping and overlaying. These development processes include:

- Rural development, closely connected to agriculture, forestry, fishery and the change of landscape through agriculture and agro-ecosystems (into cultural landscapes);
- Urban development, closely connected to industrial production, trade and commerce, administration and governance, resulting in further changes of landscapes from cultural to "techno-landscapes";
- Local (community) development that becomes connected to global development through the processes of globalisation, technical communication and action over distance;
- Population growth and demographic change processes;
- Modernisation and economic growth as societal dynamics directing development in modern society;
- Technological change and its interaction with social change processes;
- Environmental degradation and overuse of natural resources from local to global levels.

To connect complex processes in an integrated perspective that enables the formulation of strategies of sustainability governance, the regime concept is a widely used theoretical term (Holz et al. 2005). The relevant regime studies for SES include the forms of social-technical regimes (Smith et al. 2005) and socio-metabolic regimes referring to natural resource use (Krausmann et al. 2009). With the help of the mediating concept of social and ecological regimes, the abstract terms of resilience and sustainability can be translated in concrete forms of transition management that combine (in locally specific forms) the processes of

- Innovation (as creating knowledge and technologies to solve specific problems in SES and in the use of natural resources),
- Adaptation (as capacity of SES to cope with disturbance, for which innovation is a precondition), and
- Transformation (as capacity of SES, initiated by global environmental governance, to maintain long-term transformation of social and ecological systems).

Transition management in the long process of rural-urban transformation towards sustainability requires a permanent search for new possibilities and new models of development and change, for building capacities of (continuously more effective) adaptation and transformation that learn from the weaknesses of former approaches. Metropolitan areas, growing rapidly into mega-cities with many millions of inhabitants, experience worldwide similar difficulties in their efforts to transition management and local sustainable development that started after the Rio-conference in 1992. The transformation is more complicated than expected, requires long-term perspectives, new visions and social innovations, better integration of rural and urban development, new forms of cooperation of actors with different interests in transformation action groups, greater efforts and more human, social and knowledge resources than imagined. The real challenges are only gradually perceived by the actors that include governmental and non-governmental organisations, when global change affects the local development processes: for example through deterioration of the environmental conditions for agriculture or urban development through global climate change.

11.2.3 The Social and Methodological Challenges of Rural–Urban Interaction

Vulnerability studies for urban and peri-urban areas, including food security and climate change, are mainly from non-European countries, showing the practical significance of this kind of analysis as related to policies of development cooperation. The situation in Europe is specific with regard to the late phase of modernisation and post-industrial development in most countries. This has as consequence a broadening of the functions of peri-urban agriculture and land use beyond food production, including

> "the conservation of heritage landscapes, the conservation of water resources and farmland resources, and providing for both leisure and tourism activities. Anything that renders peri-urban agriculture difficult may also undermine the ability of agricultural land to support these other functions.... climate change and variability are likely to alter the capacity of these peri-urban agricultural territories to continue supporting these various functions" (Bryant et al. 2013: 60).

The multifunctionality of peri-urban agriculture can be described further through the following functions described by Zasada (2012):

- *Agricultural land-use in peri-urban areas* "contributes to the quality of life in urban regions, as it fulfils broad ranges of functions and services to the nearby urban areas".
- These functions include "food production as well as the provision of recreational services and other services related to the management of the cultural landscape, which in turn contribute to the ecological capacity of the landscape".
- *Peri-urban agriculture has two specific components*, "an intensified, high-value production on the one hand, and extensified, lifestyle and environmental-driven land-use on the other".
- *Further characteristics of peri-urban agriculture* include "(h)igh-income revenues, small-scale farm structures and the parallelism of horticulture and grassland cultivation".

- *From the perspective of farmers and land-owners* "the opportunities attached to the peri-urban framework conditions outweigh the disadvantages, which have encouraged them to adopt activities that valorise the urban demand potential".
- *In terms of planning and policy requirements* for the development of peri-urban agriculture "the main fields of action are the preservation of farmland and encouragement of multifunctional land-use, the strengthening of urban-rural relationships and the enhanced consideration and targeting of agriculture" (Zasada 2012: xiv).

The description above results from a comparison of agricultural land-use in peri-urban areas in several European countries. It does not yet show the differentiation of transition strategies and the challenges of the continuing urbanisation process and the social challenges of resilience and sustainability. Examples from case studies by this author and from other European research projects show that land use is under continuous pressure to develop innovative ideas, to adapt to social and environmental change, and to build strategies for transformation to sustainability. These strategies require experimenting and social learning from the experiences made with land use change.

The *methodological challenges* of peri-urban agriculture as part of sustainable transitions can be described as that of developing new criteria for measurement and indicators in local development in the complex processes of natural resource use that connect local and global flows of resources. These challenges are not discussed further here. We mention only some important examples for methodological tools in transition management: ecological footprints (Wackernagel and Rees 1996) to measure the land areas required for human consumption of natural resources; material and energy flow accounting (MEFA: Haberl et al. 2004) to measure the global resource flows and their inequalities; human appropriation of net primary production (HANPP: Haberl et al. 2013) to measure the share of human consumption from the primary production of ecosystems; energy return of investment (EROI: Hall et al. 2014) to measure the ratio of energy input and output in agricultural or other production processes; and planetary boundaries (Cornell 2012) to measure the global limits of natural resource use.

These indicators show different aspects of the problems of changing agricultural and other forms of production and land use, applicable also for urban and peri-urban areas. The social processes of innovation and change on the way to resilience and sustainability require improvements of resource use, also more efficient forms of conflict mitigation in natural resource use. We do not discuss these aspects further here, but show in the following illustration empirical examples from European research projects in which we participated. These examples illustrate the problems of transition management with knowledge from local case studies of peri-urban and urban agriculture and gardening.

11.3 Case Studies of Peri-Urban and Urban Agriculture and Gardening

11.3.1 Peri-Urban Agriculture

The RETHINK-project[1] researched the challenges of re-thinking farm modernisation that suffices requirements of reduced vulnerability, increased resilience and sustainable management of natural resources.

From the case studies of the RETHINK-project we summarise two studies, taken from the case study reports, from Switzerland, and Sweden that dealt with peri-urban agriculture.

The Swiss case study (Bourdin et al. 2015) in the agglomeration Bern, had as focus theme milk production in a peri-urban area and different supply chains for milk products: a dominant/conventional supply chain and new paths for milk valorisation on regional markets that include different forms, also the supply chain of organic producers. Supply chain development and management are important complementary components of peri-urban agriculture that cannot be understood from the specific conditions of production forms and processes. The logic of the two types of supply chains can be seen as similar to the differences between bio- and eco-economy strategies, the first type representing a more conventional and growth-based bio-economy, the other one an alternative "ecological" sector. An important component of the Swiss organic farming sector is the building of a national supply chain in cooperation with big retailers which makes the sector less vulnerable to minor economic shocks. The adaptability of the conventional and the organic milk producers is similarly good, and both sectors developed also transformation capacities, however, not in a coherent perspective of sustainability, rather in competing forms of "greening of agriculture". The specificities of peri-urban agriculture in the Swiss case study show that milk production and keeping of cows, sheep and goats are less difficult in peri-urban areas than keeping of pigs and poultry. Horse keeping for urban riders is found in all Swiss peri-urban areas. Altogether the changes described in the two agricultural sectors are complex and so are the processes of adaptation and transformation; this shows the necessity of developing governance structures that connect to specific networks and social learning systems. The case study argues that for the organic sector of production, because of its diversity, it is difficult to coordinate the different interests of the farmers, for example, regarding farm development. Furthermore, a contrast in interests and expectations of urban consumers and farmers is found, with the urban population often conserving a "romantic image" of farming that is not realistic with regard to the changes through farm enlargement and modernisation.

[1]RETHINK—Farm Modernisation and Rural Resilience, was a transdisciplinary research project supported by the European Commission and funding bodies in 14 countries under the umbrella of FP7 and the RURAGRI ERA-NET. For more information on the project consult the webpage http://www.rethink-net.eu/home.html

Summing up, in the Swiss case study heterogeneous trends and development processes have been identified that influence the further development of peri-urban agriculture in the perspectives of resilience and sustainability. In spite of the high adaptability agriculture in the area (as in Switzerland generally in the past decades) it is assumed that the future is not a continuity of the past agricultural development: new decreases of milk prices could, for example, have as consequence that a large part of middle size milk producers terminate their production. This will lead to the need to find other economic activities that may ensure the permanence of farms or their replacement by new forms of land use. Also, for the farms close to the city of Bern similar trends can be observed as in the Swedish case study: famers are threatened to lose farmland with the spreading of the city and new settlement and building.

In the Swedish case study (Olsson et al. 2015) agriculture in the periphery of an urban agglomeration is studied. The forms of peri-urban land use differ showing a strong influence of urban interests. The transformation of agricultural land use in the periphery of Gothenburg city confirms a growing influence of urban populations and their interests in land use on farming. In an exemplary way this can be seen in the changing forms of land use: agriculture goes away from food production, not mainly towards the new forms of bioenergy production on agricultural land, but towards extensive horse keeping. This has become an important form of land use showing the interest of urban population in riding. Horse farms are widespread in the study area, but the whole transformation of agriculture in the periphery of the city is more complex, as revealed in a longitudinal study of agricultural land use.

Today farms that use the proximity to the city for food production for the local urban market coexist with other ones that provide services for the urban population, and different forms of landscape management by farmers, also in protected areas. Four overlapping adaptive strategies of land use have been identified: (1) agricultural land use that can be changed quickly (e.g., horse farms); (2) conventional diversification and pluri-acitivity of farmers; (3) multi-functional agriculture (especially combination of food production and landscape management); (4) cereal production for different consumers.

All of these development strategies are characteristic for peri-urban agriculture under the influence of urbanization and urban interests, whereas the prior forms of small-scale and mixed agriculture for local markets have vanished. They represented the last form of a conventional agriculture that developed within the national Swedish agricultural modernisation policy after the Second World War, already influenced by urban markets. With the growing influence of urbanisation on farming adaptability and transformability as requirements of social-ecological resilience became more important for farmers. The development of farms appears as less stable in the long run; farmers need to adapt and transform their agriculture continuously and more actively, trying to find new forms to be able to continue farming.

The development of peri-urban agriculture showed two phases: first the transformation of small family farms to diverse new forms of farming, and in the second phase additional forms of differentiation of agriculture under the influence of urbanisation-driven change, with two dominant forms: food production for the local urban market and horse keeping.

The long-term trends of land use change identified in the case study include: agricultural land is transformed into urban land for building (presently minority); arable land transformed to other use, livestock grazing and riding (majority); continuing agricultural food production (minority); abandoning of livestock grazing in the outlands that transform into new forest areas. This last form is specific for the metropolitan area of Gothenburg, a trend that differs strongly from other metropolitan areas in Europe: large areas in the urban periphery were no longer used for agriculture, but reforested.

Complementary to the phenomena of agricultural transformation the following development forms influencing peri-urban agriculture are important: large parts of peri-urban agricultural land belong administratively to the city; it is a coastal area with competitive use of land for the urban (industrial) and third sector economy (transport and communication, e.g. harbours, local and supra-local tourism, seasonal dwelling and commuting of urban residents, land use for sports and recreation). Collaboration among farmers developed in this area since long time in specific forms of agricultural modernisation (cooperatives, the early phase) and the general forms of local, community-based cooperation that included also agriculture (local movements, with active support through governmental institutions). The local movement- and network-based, often informal, cooperation is still influential in the late-modern peri-urban development of agriculture, whereas direct cooperation of farmers (e.g. through machine rings) has become less important.

In the Swedish case study resilience and sustainability in the urban fringe are developing through a culture of social learning that supports the adaptation and transformation of peri-urban agriculture and shows the blending of rural and urban traditions of communication: community-based local movements, urbanisation of the countryside, and the inclusion of land use planning into urban planning that includes agricultural land and protected areas in the urban periphery.

Further case-studies in other European countries participating in the RETHINK-project dealt with agricultural transition to resilience and sustainability in various forms of rural areas. From all the case studies (accessible through the project website RETHINK) the requirements of adaptability and transformability of agricultural land use can be described as follows:

1. *Matching the contrasting requirements of permanence and change* is a general requirement of resilience and sustainability for all forms of agriculture studied, in a process perspective where relative stability/persistence is achieved through adaptation of farms to changing conditions (continuous process), and at certain times through transformation (transition to other production systems, far reaching system changes, rupture of development paths).
2. *Matching autonomy* (as enabling change) and *network embeddedness* (enabling efficiency and providing information) of farms is a precondition for resilience and sustainability transformation where farmers become participants in larger development-directing forms and networks of cooperation.
3. *Unfolding cooperation* that is supported through local networks and movements is often seen as contrasting with the power- based hierarchies created by

governmental organisations. But integration of top-down and bottom-up perspectives in sustainability governance becomes a main requirement of future development.

4. *Informal social networks* are supporting the resilience- and sustainability-oriented innovations, adaptation and transformation of farming, in the case studies mainly illustrated through organic farming and other forms of environment-friendly agricultural production.
5. *Social learning:* favourable conditions and contexts for social learning that support the development of social-ecological resilience include a variety of factors that create resilience—cooperation between farmers and across sectors, social networks that include other actors than farmers, development and change of farms that allow for adaptation disturbances, shocks and changing conditions of markets and environmental conditions in the longer perspective of sustainability transformation.
6. *Diversity of production forms* and activities at the farms in the study areas (in difference to diversification of the production and other income-generating activities on the single farm) is a context component that may support resilience and sustainability regarding social and ecological diversity at landscape or regional levels. However, it includes also contrasting factors that do not support resilience, being often mainly market-oriented adaptation in the short run.
7. *Resistance to change* can be found in strategies ignoring resilience and sustainability in attempts to continue agricultural development on specialisation and growth based development paths. Such resistance to change is often connected with the orientation of farmers to the conventional logic of modernisation, to food production and to growth that contrasts with transformation processes towards forms of agriculture compatible with the criteria of an eco-economy or sustainability.
8. *Threats and tensions* emerging in the processes of adaptation and transformation of agriculture include for farms the insecurity about the long-term future of agriculture, the need of high investments, the high workloads to deal with the bureaucratic requirements of regulation and policy, the high prices of land, and the competing land use demands from other economic sectors and urbanization. From other sectors of the regional economy and from local inhabitants agriculture does not always get sufficient support.

The adaptation and transformation processes, confronted with these contrasting requirements, do not just require technical and social innovations. Innovations can only become effective when they are combined with other capacities, for example in peri-urban agriculture with that of flexibility in land use as it is described in exemplary forms as multifunctional agriculture (Renting et al. 2009).

Experiences in European countries with policies and strategies of adaptation and transition to sustainability in connected rural and urban areas show the growing importance and the differentiating forms of urban agriculture and gardening. In the following section we describe an example of an innovative project of urban gardening that shows as well the difficulties as the possibilities of transition to sustainability.

11.3.2 Project "Urban Gardening": Case Study Lisbon

In the Portuguese case study in Lisbon urban gardening is at the same time contributing to reinforce biodiversity, to increase the resilience of the city to floods, and contributing to increase family income of immigrant communities and families hit during the recent economic crisis[2]. Allotments started to grow spontaneously since the early 1960's mainly in the peri-urban areas, related first with the migration from rural areas to the city, and, in the 1980's associated with migrant communities, especially those coming from former Portuguese colonies (Cabannes and Raposo 2013; Matos and Batista 2013; Cabral 2014). More recently the economic crisis and high unemployment rates have transformed this small farming production into a fundamental mean of subsistence for many families. The newcomers in the city started using and occupying urban voids, both municipal and private.

Therefore, the Lisbon municipality decided to intervene, planning and integrating the spontaneous "movement activities" into the development of green infrastructure of the city. This process of reorganisation of non-regulated allotments is part of the Lisbon's Green Plan, adopted in 2007, were agriculture was assumed as an important component of the ecological structure of the city; urban agriculture provides not only food but several ecosystem services that are essential to establish a green infrastructure and to connect urban, peri-urban and rural areas functionally (CML 2016). Within this framework a Strategy for Urban Agriculture was defined and the Municipality started a process of regulation of these areas creating Horticulture Parks. They consist of urban infrastructures used by farmers, but they are also open to the public, for different leisure-time uses and creating pathways for pedestrians and bicycles, approaching the agricultural activities of the remaining population. These horticultural parks aimed at addressing several challenges: (a) to enlarge the scarce green spaces in the city; (b) to link most green spaces through ecological corridors; (c) to mitigate the impact of channelling waterlines by creating water basins in strategic valleys and to provide ecological services; (d) to reorganise allotments that were growing fast (partly due to economic crisis and rising unemployment) and unorganised; (e) to meet the increasing demand for allotments driven by middle income family's desire to establish a healthy life style, to connect to nature and to ensure the quality of food products they eat, and, (f) to contribute to food production providing quality food (organic production is mandatory or highly incentivised) (Matos and Batista 2013; Bernardo 2013). A total of 20 horticulture parks (municipal allotments) were projected to be created until 2017; in 2014 ten horticultural parks were already open for the public (CML 2016).

---

[2]See, furthermore, in this book, the chapter from Fassi and Sedini, discussing an interesting case COLTIVANDO —The convivial garden at the Politecnico di Milano. The recently published book by Calori and Magarini (2015) gives examples on sustainable food policies from more cities; our text describes the processes of transformation to sustainability in broader terms, as more complex social processes from which the food policies are only a part.

This process implied reorganising and unifying the plots, providing infrastructures, water access, and small sheds for tools storage. Two types of allotments were created varying in size and function:

1. the plot located in a social allotment park has around 150 square meters, is meant for subsistence and the surpluses can be sold;
2. a recreational or pedagogic allotment park holds in average plots of 100 square meters, organic production is mandatory and is only for own consumption.

The policy intends to respond to the increasing number of families who wanted to "return to earth" and produce their own organic food. In both cases an annual fee is due as a contribution to maintenance, technical training on organic production and water use, but low income families get discounts that can reach 80%. Due to the increasing demand for allotments, the city has not only organized the former allotments but also created new ones.

Among those horticultural parks Chelas Valey is the largest, covering about 15 hectares, of which 6.5 are used for urban gardens, including 400 plots, each with 150 square meters, with a share allocated directly to about 100 people who had already unofficially created allotments (Cabannes and Raposo 2013), with the remaining reserved for a public tender carried out in 2013. Today in those plots coexist "old farmers" in activity for several decades and others who only started after the recent reorganization of the allotments. The relationship among them is reported to be very good, conflicts are rare. Collaboration is a norm; they exchange agricultural practices, seeds and products. The type of crops grown is diversified according to the nationality of the farmers from Portugal, India, Cape Verde and Angola (Luz and Pires 2014).

The process of developing these horticultural parks was top-down, totally designed by the Municipality. The previous users were not consulted or involved in the process, they have just been notified that they should leave the place during rehabilitation works and could return later. But they had to comply with the rules defined by Municipality, namely: the organic mode of production (after technical training offered by the municipality), composting, no use of chemical pesticides or herbicides, of chemical fertilizers, of Genetic Modified Organisms (GMO), of infesting species, and no construction of unauthorizsd types of fences or shelter structures, or planting of trees.

Nevertheless, and although still ongoing, the reorganisation process of informal allotments seems to have turned into a positive and cooperative strategy for urban transformation. Changes that came with the municipal intervention were perceived as stimulating by the older farmers that emphasised access to water and other infrastructures as an improvement justifying the introduction of annual fees. At the same time the development attracted new users for leisure activities or new urban farmers (Luz and Pires 2014).

In Lisbon urban gardening is contributing to increase resilience in the double sense of (a) resilience of the city by improving its capacity of producing food and providing ecological services, as well as (b) resilience of families and individuals towards economic crises. At the same time urban gardening also provides a momentum for increasing social cohesion and integration of immigrants. In the social allotments the users report positive impacts in socialisation (of those who are retired), food security of the family (in the case of unemployed), and even creating opportunities for small scale entrepreneurship as they are allowed to sell surpluses. It seems that this project under the guidance of the municipality has created an innovative social infrastructure for transformation to sustainability on which further and more large-scale projects of sustainability governance can build (Luz and Pires 2014).

11.4 Discussion: Integrated Local Strategies for Innovation and Socio-ecological Transformation

Local strategies for innovation and socio-ecological transformation need to work with contradicting and contrasting requirements as a continuous challenge. This can also be described as requirement of inclusive and multi scale politics or governance processes, as in the commentary article on "transforming innovation for sustainability" by Leach et al. (2012). These authors formulate the framing conditions and perspectives that allow further discussion of the examples of peri-urban and urban agricultural projects described above. The authors summarise their reflections as follows:

> "The urgency of charting pathways to sustainability that keep human societies within a 'safe operating space' has now been clarified. Crises in climate, food, biodiversity, and energy are already playing out across local and global scales and are set to increase as we approach critical thresholds. … ambitious Sustainable Development Goals are now required along with major transformation, not only in policies and technologies, but in modes of innovation themselves, to meet them. … such 'transformative innovation' needs to give far greater recognition and power to grassroots innovation actors and processes, involving them within an inclusive, multi-scale innovation politics. The three dimensions of direction, diversity, and distribution along with new forms of 'sustainability brokering' can help guide the kinds of analysis and decision making now needed to safeguard our planet for current and future generations." (Leach et al. 2012:1).

What the authors describe as "radically new approach to innovation" includes the following components:

1. Re-directing of change in accordance with criteria of sustainability,
2. Supporting diversity and experimenting with different approaches of policy innovation,
3. Distribution in the sense of sharing the burdens and the advantages from transformation.

These are three procedural requirements that can also be applied in the examples we described. The challenge described by the authors as connecting local and grassroots innovation capacity with the requirements of global change and planetary boundaries of resource use (Leach et al. 2012: 5) can be seen as necessity of all strategies of sustainability governance.

Important common elements in the varying conditions for local, urban-rural projects for resilience and sustainability can be seen in the capacities to learn and to cooperate that are required from the heterogeneous actors participating the processes of local development and transformation. These processes of social learning and cooperation of actors reflect the complexity and elasticity of resilience and sustainability that are seen as examples of "essentially contested concepts" (Collier et al. 2006, see above: conceptual challenges of transition strategies). Both of the concepts are defined and interpreted in many different ways, no consensus about their interpretation is achieved; but still they can be applied in meaningful ways, as we tried to show. Furthermore, the concepts are necessary to deal with the global environmental problems. It can be argued, that resilience and sustainability became overused and over-interpreted, give no longer a safe compass and guidance for sustainable transformations. But this seems more to show the nature of the problems to deal with than the bad quality of the concepts. The future sustainable society is unknown and has to be built with knowledge developed and experience gained in the transformation processes. Transformation needs to be constructed anew and continually adapted in the long process, with measurable concepts. In this transformation process the ideas of resilience and sustainability require re-interpretation and modification with the growth of scientific knowledge and the changing conditions and consequences of global change that affect local development in unforeseeable ways. For cities the unforeseeable future can be seen as a paradox that includes the urban sustainability multiplier described by Rees: cities do not have ecological disadvantages only, also a series of advantages that help to save material, energy, space and using resources effectively, thus chances for more sustainable resource use. These contrasting qualities of cities stimulate social and technical innovations in search of a future sustainability; furthermore, the contrasts make such strategies of sustainability governance as the development of new forms of urban and peri-urban agriculture a necessity of further rural-urban development. The contrasts of rural and urban areas, of rural and urban development, of hinterland and global cities, have now reached the cities themselves that reconnect rurality and urbanity in the urban landscapes.

11.5 Conclusions: Requirements of Further Development of Integrated Transformation Strategies

Integrated urban-rural sustainable development requires new governance models for effective adaptation and transformation; and it requires learning from the weaknesses of former approaches. Metropolitan areas experience worldwide similar difficulties, paradoxes and challenges in their efforts of transition management and local sustainable development after the Rio-conference in 1992. The transformation processes are more complicated than expected; they require long-term perspectives, new visions and social innovations, better integration of rural and urban development, greater efforts and more human, social and knowledge resources than imagined. The real challenges are only gradually perceived, when global change affects local development. From the examples we described and discussed, we can derive the main requirements of integrated strategies of resilience and sustainability as follows:

- Achieving sustainability implies more complex and systematic forms of inter- and transdisciplinary knowledge integration; these include integration of scientific knowledge from the social and natural sciences and practical, for example local ecological knowledge, from social actors and practitioners.
- Furthermore, specific forms of collective and social learning by the actors involved, e.g., "double loop" learning are required. This implies not only learning to develop joint action strategies, but the learning to anticipate and take into account in present action its future consequences. The challenges of such social learning include that of dealing with complexity, uncertainty, conflicts and power asymmetries.
- Strategies to enhance the transformative capacity of the SES include different components—strengthening collective action and cooperation of resource users; developing mechanisms of multi-scale and multi-actor governance to deal with contrasting requirements; building transformative capacity of individuals, groups and institutions; developing process models of navigating transformations through different stages of development, with periods of turbulence and uncertainty (Olsson et al. 2006).
- Transformation networks are networks of social actors and institutions that are able to initiate and maintain processes of socio-ecological transformation. They are a core component of governance for sustainability transformations, at different levels and scales of action. The capacity of such networks implies more than political action and coordination: complex forms of collective action in which social, cultural, political, economic and ecological changes are integrated in the broader processes of socio-ecological transformation.

In all processes of sustainability transformation time is a key aspect, and an "unknown variable": the future is open and unknown; it is not determined by our present action, but influenced in ways we cannot foresee. Transformations to sustainability require long-term perspectives—of several generations or centuries—and for such a long process no forms of action can be kept during the whole process. This underlines the necessities of experimenting and the capacities of social learning and cooperation, to realise the changes of strategies that are necessary in the transformation process.

References

Adger, N. (2000). Social and ecological resilience: Are they related? *Progress in Human Geography, 24*(3), 347–364.

Armitage, D., & Plummer, R. (Eds.). (2010). *Adaptive capacity and environmental governance*. Heidelberg: Springer.

Asara, V., Otero, I., Demaria, F., & Corbera, E. (2015). Socially sustainable degrowth as a social-ecological transformation: Repoliticizing sustainability. *Sustainability Science*. https://doi.org/10.1007/s11625-015-0321-9

Beck, U. (1986). *Risikogesellschaft: Auf dem Weg in eine Andere Moderne*. Frankfurt a M: Suhrkamp.

Bernardo, J. M. (2013). *Campo na Cidade: As Hortas e os Hortelões de Lisboa*, In Paisagem e Património - Aproximações Multidisciplinares. Equações de Arquitectura, Dafne Editora.

Bourdin, D., Gertz, A., Réviron, S., & Siegenthaler, M. (2015). *Sub-urban food production in a Swiss agglomeration.* Case Study Report, Rethink-project, AGRIDEA – Swiss Association for the Development of Agriculture and Rural Areas, Lausanne/Lindau.

Bruckmeier, K. (2013). *Natural resource use and global change: New perspectives in social Ecolog*. Basingstoke: Palgrave Macmillan.

Bruckmeier, K. (2016). *Social-ecological transformation: Reconnecting society and nature*. London: Palgrave Macmillan.

Bruckmeier, K., & Olsson, G. (2014). *An attempt to clarify the resilience concept for renewed strategies of farm modernization.* In A. T. Aenis et al. (Eds.), 11th European IFSA Symposium, Farming Systems Facing Global Challenges: Capacities and Strategies, Berlin, Germany, 2–4 April 2014, Proceedings.

Bryant, C., Carvajal Sánchez, N., Delusca, K., Daouda, O., & Sarr, A. (2013). Metropolitan vulnerability and strategic roles for Periurban agricultural territories in the context of climate change and variability. *Cuadernos de Geografia, Revista Colombiana de Geografia, 22*(2), 55–68.

Cabannes, Y., & Raposo, I. (2013). Peri-urban agriculture, social inclusion of migrant population, and right to the City. *City, 17*(2), 235–250.

Cabell, J. F., & Oelofse, M. (2012). An Indicator system for assessing Agroecosystem resilience. *Ecology and Society, 17*(1), 18.

Cabral, M. I. (2014). *Urban Gardening in Leipzig and Lisbon*. Scientific Report COST ACTION TU1201. Retrieved from http://www.urbanallotments.eu/fileadmin/uag/media/STSM/STSM_report_Cabral_Leipzig_Lisbon_dec_2014.pdf

Calori, A., & Magarini, A. (2015). *Food and the cities: Food policy for sustainable cities*. Milano: Edizioni Ambiente.

Chapin, F. S., III, Kofinas, G. P., & Folke, C. (Eds.). (2009). *Principles of ecosystem stewardship: Resilience-based natural resource Management in a Changing World.* New York: Springer.

CML. (2016). *Parques Hortícolas Municipais*. Retrieved February 2016, from http://www.cm-lisboa.pt/viver/ambiente/parques-horticolas-municipais.

Collier, D., Hidalgo, F. D., & Maciuceanu, A. O. (2006). Essentially contested concepts: Debates and applications. *Journal of Political Ideologies, 11*(13), 211–246.

Cornell, S. (2012). 'O' the system properties of the planetary boundaries. *Ecology and Society, 17* (1), 2. https://doi.org/10.5751/ES-04731-1701r02

Cutter, S. L., Boruff, B. J., & Shirley, W. (2003). Social vulnerability to environmental hazards. *Social Science Quarterly, 84*(2), 242–261.

Diekmann, A., & Preisendörfer, P. (2001). *Umweltsoziologie. Eine Einführung*. Reinbek: Rowohlt.

Fischer Kowalski, M., & Haberl, H. (Eds.). (2007). *Socioecological transitions and global change. Trajectories of social metabolism and land use*. Cheltenham: Edward Elgar.

Fischer-Kowalski, M., & Rotmans, J. (2009). Conceptualizing, observing, and influencing social-ecological transitions. *Ecology and Society, 14*(2), 3. Retrieved from http://www.ecologyandsociety.org/vol4/iss2/art3/

Folke, C. (2006). Resilience: The emergence of a perspective for social-ecological systems analysis. *Global Environmental Change, 16*, 253–267.

Folke, C., Hahn, T., Olsson, P., & Norberg, J. (2005). Adaptive governance of social-ecological systems. *Annual Review of Environment and Resources, 30*, 441–473.

Füssel, H. M. (2007). Vulnerability: A generally applicable conceptual framework for climate change research. *Global Environmental Change, 17*, 155–167.

Gunderson, L. H., & Holling, C. S. (Eds.). (2002). *Panarchy: Understanding transformations in human and natural systems*. Washington, DC: Island Press.
Haberl, H., Fischer-Kowalski, M., Krausmann, F., Weisz, H., & Winiwarter, V. (2004). Progress towards sustainability? What the conceptual framework of material and energy flow accounting (MEFA) can offer. *Land Use Policy, 21*, 199–213.
Haberl, H., Erb, K. H., & Krausmann, F. (2013). Global human appropriation of net primary production (HANPP). *The Encyclopedia of Earth.* Retrieved from http://www.eoearth.org/view/article/153031.
Hall, C. A. S., Lambert, J. G., & Balogh, S. B. (2014). EROI of different fuels and the implications for society. *Energy Policy, 64*, 141–152.
Holz, G., Brugnach, M., & Pahl-Wostl, C. (2005). Specifying "regime": A framework for defining and describing regimes in transition research. *Technological Forecasting and Social Change, 75* (5), 623–643.
King, C. A. (2008). Community resilience and contemporary Agri-ecological systems: Reconnecting people and food, and people with people. *Systems Research and Behavioural Science, 25*, 111–124.
Knight, F. H. (1921). *Risk, uncertainty, and profit.* Boston: Hart.
Krausmann, F., Fischer-Kowalski, M., Schandl, H., & Eisenmenger, N. (2009). The global socio-metabolic transition: Past and present metabolic profiles and their future trajectories. *Journal of Industrial Ecology, 12*(5–6), 637–656.
Leach, M. J., Rockström, P., Raskin, I., Scoones, A. C., Stirling, A., Smith, J., et al. (2012). Transforming innovation for sustainability. *Ecology and Society, 17*(2), 11. https://doi.org/10.5751/ES-04933-170211
Lloyd, M. G., Peel, D., & Duck, R. W. (2013). Towards a social-ecological resilience framework for coastal planning. *Land Use Policy, 30*, 925–933.
Luhmann, N. (1991). *Soziologie des Risikos*. Berlin: De Gruyter.
Lundgren, L., & Jonsson, A. (2012). *Assessment of social vulnerability: A literature review of vulnerability related to climate change and natural hazards (CSPR briefing 9)*. Linköping: Linköping University Centre for Climate Science and Policy Research.
Luz, A., & Pires, I. (2014). *Regularização das Hortas Urbanas na Cidade de Lisboa - as Hortas Sociais do Vale de Chelas.* In I. C. da Silva, M. Pignatelli & S. de M. Viegas (coord.), Livro de Atas do 1° Congresso da Associação Internacional de Ciências Sociais e Humanas em Língua Portuguesa, 6935–6952. ISBN: 978–989–99357-0-9.
Markard, J., Raven, B., & Truffer, B. (2012). Sustainability transitions: An emerging field of research and its prospects. *Research Policy, 41*, 955–967.
Matos, R., & Batista, D. (2013). Urban agriculture: The allotment gardens. In *Advances in landscape architecture*. Rijeka: INTECH open science/open minds.
Mol, A. P. J., Sonnenfeld, D. A., & Spaargaren, G. (Eds.). (2009). *The ecological modernisation reader: Environment*. London: Routledge.
Olsson, P., Gunderson, L. H., Carpenter, S. R., Ryan, P., Lebel, L., Folke, C., & Holling, C. S. (2006). Shooting the rapids: Navigating transitions to adaptive governance of social-ecological systems. *Ecology and Society, 11*(1), 18. http://www.ecologyandsociety.org/vol11/iss1/art18/
Olsson, G. A., Bruckmeier, K., & Wästfelt, A. (2015). *Peri-urban agricultural transformations in Gothenburg, Sweden*. Case study report, RETHINK-project. Gothenburg, Sweden: Gothenburg University, School of Global Studies.
Paavola, J., & Hubacek, K. (2013). Ecosystem services, governance, and stakeholder participation: An introduction. *Ecology and Society, 18*(4), 42.
Raskin, P. D., Electris, C., & Rosen, R. A. (2010). The century ahead: Searching for sustainability. *Sustainability, 3*(8), 2626–2651.
Renn, O., & Klinke, A. (2004). systemic risks: A new challenge for risk management. *EMBO Reports, 5*(suppl. 1), 41–46.

Renting, H., Rossing, W. A. H., Groot, J. C. J., Van der Ploeg, J. D., Laurent, C., Perraud, D., et al. (2009). Exploring multifunctional agriculture. A review of conceptual approaches and prospects for an integrative transitional framework. *Journal of Environmental Management, 90*, 112–123.
Scoones, I. (2009). Livelihoods perspectives and rural development. *Journal of Peasant Studies, 36* (1), 1–26.
Smith, A., Stirling, A., & Berkhout, F. (2005). The governance of sustainable socio-technical transitions. *Research Policy, 34*, 1491–1510.
Wackernagel, M., & Rees, W. E. (1996). *Our ecological footprint: Reducing human impact on the earth*. British Columbia: Gabriola Press New Society Publishing.
Zasada, I. (2012). *Peri-urban Agriculture and Multifunctionality: Urban Influence, Farm Adaptation Behaviour and Development Perspective*. Dissertation, Technical University Munich.

Chapter 12
Territorial Innovation Models: Which Consequences in Terms of Policy Design for Peripheral Regions? A Portuguese Perspective

Domingos Santos

12.1 Introduction

In the last three decades, innovation has become broadly understood "to include product, process and organizational innovation in the firm as well as social and institutional innovation at the level of an industry, region and nation" (Morgan 1997: 492) and is a critical dimension in the analysis of territorial development.

As innovation processes have inherently a strong territorial and social matrix, then it must be emphasized the progressively prominence that an enlarged set of features now assume in the production of knowledge for innovation, namely the informal contacts and the flows of tacit knowledge amongst the different type of actors, their conventional rules and cultural patterns (Storper and Scott 1995), their relational capital and their social capital, on the sense proposed by Putnam (1993: 35): "features of social organization, such as networks, norms and trust that facilitate coordination and co-operation for mutual benefit". There has been a shift towards the understanding of the innovation process as a socially built mechanism based on the accumulation of knowledge (codified or tacit) through a continuous and collaborative learning course (Lawson and Lorenz 1999; Tura and Harmaakorpi 2005). Accordingly, Maskell and Malmberg (1999: 20) argue that, more than ever before, territorial competitiveness is now concerned with "knowledge creation and with the development of localized capabilities that promote learning processes".

In this sense, the dynamics of innovation is based on resources that are place-specific; so, regionally based complexes of innovation and production are increasingly the privileged instruments to harness and recreate knowledge and intelligence.

D. Santos (✉)
Polytechnic Institute Castelo Branco, Castelo Branco, Portugal

CICS.NOVA – Interdisciplinary Centre of Social Sciences, Lisbon, Portugal
e-mail: domingos.santos@ipcb.pt

H. Pinto et al. (eds.), *Resilience and Regional Dynamics*, Advances in Spatial Science, https://doi.org/10.1007/978-3-319-95135-5_12

The accumulated knowledge that production systems develop because they are incorporated in locally based institutions and, in a generally non-mobile workforce, tend to prolong competitive advantages, however, while proximity does matter, what really seems critical for the upgrading of the competitive edge of localized production systems and resource creation is in fact organizational proximity (Fujita and Krugman 2004; Asheim and Coenen 2005; Carlsson 2005; Shearmur 2011). It is therefore important to recognize that "knowledge transmission and collective learning may be nurtured by cultural, institutional and geographical proximities often in combination" (Keeble and Wilkinson 1999: 300).

So, on the last three decades, there has clearly been a change of paradigm on the perception of the relation between industrial dynamics and regional development: long-term regional competitiveness and sustainability has less to do with cost-efficiency and more to do with the ability of firms and institutions to innovate, or, in broader terms, to upgrade their knowledge base.

The academic discussion about the dialectics innovation-territory remains mostly at the abstract and theoretical level. As a result, a significant operationalization of key concepts is needed in order to enhance the empirical investigation (Moulaert and Sekia 2003). The repercussions of this problematic on least favored regions have seldom been analyzed. Usually, the analysis is focused on urban-metropolitan areas on medium to high-tech sectors. The knowledge provided by the approaches that analyse the dialectics innovation-territory is very enriching and gives new insights about possible policy interventions in peripheral regions.

12.2 Innovation and Territory: The Analytical Framework

It is claimed that regional dynamics produce idiosyncratic interdependences amid the regional stakeholders that develop into a specific economic and technological course. Some theoretical and methodological frameworks converge on this perspective, namely the *Industrial District* approach, the *Innovative Milieu* paradigm, the *Learning Region* approach and the *Regional Innovation Systems* model.

The concept of industrial district, affiliated on a Marshallian analysis of socio-economic organization and on the notion of agglomeration externalities, is noticeably rooted on the research about the *Third Italy* and authors like Bagnasco, Garofoli and Becattini. The concept is related to a high concentration of horizontally integrated, specialized and autonomous SMEs, each one related to a distinct stage of manufacture. These small enterprises cooperate actively to create a wide sort of differentiated goods that are sold on customer-oriented, disjointed and varied global markets. The local economies often take advantage of the information exchange made possible by the progress of localized producer-user networks, following the flattening of vertical integration within firms (Torre and Wallet 2014).

The following factors are highlighted as the factual sources of regional dynamics and competitiveness, as Cappello (1996: 488) refers: "entrepreneurship, production flexibility, district economies and the presence of some *collective agents* capable of acting as a catalyst for the mobilization of the indigenous potential (a local bank, wholesalers, local industrial associations, etc.)". This local complex of firms is densely inter-connected by a social division of labor (Morrison 2008). Storper (1995) accentuates the role of localized *untraded interdependencies* (labor market, local conventions, etc.).

In 1985, the Groupe de Recherche Européen sur les Milieux Innovateurs (GREMI) proposed a theoretical approach founded on the decrease of transaction costs, but also on the innovative dynamics resulting from territorial externalities. The approach is grounded on the existence of innovation network that vertebrate the territories, simultaneously cause and effect of the collective and interactive nature of the innovation process.

Maillat (1998: 124) establishes a useful distinction: "the innovative *milieu* is not a specific category of localized production system but a cognitive set . . . (it) corresponds to a territorialized, outwardly open complex, that is, open to technological and market environment, which incorporates and masters know-how, rules and relational capital." Innovation is understood as the incorporation of information and resources by the territory, therefore largely exceeding the simplistic meaning of innovation as a purely technological output.

This constitutes the most interesting value-added in comparison to the industrial district model: innovation intrinsically has both a territorial and institutional dimension. The promotion of local and regional synergies is the driver of the innovative milieu—in other words, the territory is understood simultaneously as cause and consequence of the stakeholders' cooperative behaviors and their learning dynamics.

The *Learning Region* approach accompanied the innovative milieu model, proposing a similar perspective and widening its ambit to the ICT challenges and opportunities. This approach has been worked principally by Scandinavian (Lundvall, Asheim, Isaksen) Welsh (Cooke and Morgan), concentrating on two analytical dimensions:

- on the one hand, the strengthening of the relational perspective: as the innovation dynamics requires continues access to flows of data, information and knowledge, the capability to innovate requires a networking strategy;
- on the other hand, it highlights the increasing significance of processes of information and knowledge creation, dissemination and absorption; it is now almost a refrain the very often quoted Lundvall's (1992) statement that "knowledge is the most fundamental resource and learning the most important process" and thereby the territory must adopt a framework conducive to knowledge production and learning.

As Ferrão (1997) proposes, the learning region approach emphasizes the centrality of the collective learning mechanisms while levers of regional competitiveness and sustainability. Overall, the learning region and the innovative milieu approaches possess clear similarities, the former being visibly a semantic derivation of the approach that seems more elaborated and structured. That is the motive why on Table 12.1 there is no difference concerning these two approaches (Santos 2009).

Table 12.1 Industrial district, innovative milieu/learning region and regional innovation system: a synthesis. Source: Santos (2009)

	Industrial district	Innovative milieu/ learning region	Regional innovation system
Emergence	Spontaneous; as local productive system	Spontaneous/induced; as cognitive entity	Induced; as organizational entity
Predominant *culture*	Industrial atmosphere	Entrepreneurial culture	Scientific and entrepreneurial culture
Productive system	Industrial; productive specialization; specialization in line with a sectoral division of labor; SMEs; vertically disintegrated; self-centered	Industrial and tertiary; diversification of production in terms of intra-industry division of labor; large and SMEs; quasi-vertical integration; open	Industrial and tertiary; diversification of production from the standpoint of intra-industry division of labor; large and SMEs; quasi-vertical integration; open
Non-mercantile relations among the firms	High intensity of extra-productive exchanges; informal inter-personal networks of information flows; strong horizontal and vertical mobility of labor	High intensity of extra-productive exchanges; diversity of non-market formal relations	High intensity of extra-productive exchanges; diversity of non-market formal relations
External relations	Open to the outside world through suppliers and clients	Open to the outside; insertion on the international circuits of information and knowledge transfer	Strong opening to the outside; insertion on the international circuits of information and knowledge transfer
Reticular structures	Compacts; networks without a strategic center	Compacts; networks with leader enterprises or with pivot enterprises	Networks with pivot enterprises or institutions (university, . . .)
Logics	Communitarian; of survival; to avoid that the regional economies act as mere spaces of localization of exogenous investments	Of partnership; creation of collective learning mechanisms as instruments of the competitive renewal of the productive basis	Of partnership; institutional architecture as a lever of the territorial competitiveness; promotion of the innovation potential
Dominant forms of knowledge	Tacit; contextual	Codified; global	Codified; global
Dominant forms of learning	*By doing, by using, by interacting*	*By doing, by interacting, by networking*	*By searching, by networking*
Dominant modalities of innovation	Incremental; adaptive of the product and of the process	Incremental and radical—*first of its* kind; of the product, of the process and organizational	Incremental and radical—*first of its* kind; of the product, of the process and organizational

(continued)

Table 12.1 (continued)

	Industrial district	Innovative milieu/ learning region	Regional innovation system
Growth dynamics	Competition-emulation-cooperation; based on an enlarged social mobilization; entrepreneurial risk socially supported	Competition-cooperation; induced by the activation of knowledge flows; entrepreneurial risk institutionally supported	Cross-fertilization; highly induced by the institutional universe; dynamic adjustment between the entrepreneurial and the institutional spheres
Potential risks	Socio-technological lock-in; barriers to the entrance of new players; growth of firm "hierarchisation" phenomenon; deviant behaviors	Technological and relational lock-in; exit barriers	Technological and relational lock-in; exit barriers: Institutional sclerosis

Innovation is the product of multi-level networking of flows of information and knowledge (Cooke 1996; Morgan 1997; Cooke et al. 2005; Tura and Harmaakorpi 2005; Cooke 2008). In a knowledge-intensive territory, intellectual competencies replace physical labor as the critical dimension of value creation and tool for increased competitive advantage.

A comprehensive meaning of the innovation system comprises not only R&D institutions but also the productive fabric, its institutional and governance base, its financial configuration and its educational and training facilities. Such a system can therefore be defined as a specific format of organization and regulation of the stakeholders' relations through the innovative and co-creative dynamics. Different territories can display distinct or idiosyncratic systems of innovation which depart from the national norm and in turn be different from other regions (Bair 2008; Balland et al. 2015).

Thus, it appears convenient to distinguish two diverse configurations of regional innovation systems, as Asheim et al. suggest:

> on the one hand, we find innovation systems that are parts of a regionalized national innovation system, i.e. parts of the production structure and the institutional infrastructure located in a region but functionally integrated in, or equivalent to, national (or international) innovation systems, which is based on a top-down, linear model of innovation. On the other hand, we can identify innovation systems constituted by the parts of the production structure and institutional set-up that is territorially integrated or embedded within a particular region, and built up by a bottom-up, interactive innovation model.

It seems important to examine the innovation dynamics through this bottom-up, territorial methodological angle, as suggested by the innovative *milieu*, the learning regions or the regional systems conceptual models (Cooke 1996; Asheim et al. 2011), a complementary lens of the functional and sectoral methodology, allowing to capture the flows and the mode interaction occurs at territorial level.

The pluralism of interpretations of innovation dynamics converges, however, on the understanding of the importance of the collective learning processes, networking and governance. More profound and lasting effects of increased competitiveness can only be obtained if innovation becomes systemic in a region—that is, if it assumes a territorial innovation system configuration.

This debate about the nature of innovation and its implications at a territorial level has led to the gradual recognition that innovation is neither a one-way diffusion process, nor a clear-cut factor-impact relationship between the creative innovative entrepreneur and the firm, but a process and/or a system.

12.3 The Portuguese Context: A Brief Overview

Some research studies conducted in different areas of Portugal, such as the Península of Setúbal (Almeida 1994), the district of Aveiro (CEC 1997), Alcanena (Nicolau 2001), the Northern region (Mota Campos and Silva 1997), the Urban Arch of the Interior Centre of Portugal (an area involving the municipalities of Castelo Branco, Fundão, Covilhã and Belmonte, about the textile-clothing industry) (Santos 2012), the Pinhal Interior Sul (a rural area on the Centro Region, around the firms of the wood *filière*) (Santos and Simões 2008) and on different digital regions (Simões 2008) have been underlining the weak interactiveness amongst the territorial stakeholders, an unfavorable context that largely constrains the regional innovation potential.

12.3.1 A Highly Concentrated National Innovation System

The Portuguese S&T system is comparatively weak in European terms. The percentage of R&D expenditure in GDP in 2013 was only 1.33%, being the responsibility of universities and other public research institutions (57.7%). The industry has been augmenting its weight very quickly, mainly in technological intensive activities, the bulk of R&D expenditures at this level being concentrated in a small number of sectors and companies (Table 12.2).

There are a vast number of R&D institutions with a good scientific status and very qualified human resources, however, the mechanisms of knowledge transfer to industry still lack effectiveness and continuity.

It is worth adding that the Portuguese S&T system is territorially very asymmetric (Table 12.3), with a disproportionate concentration of resourced in the Lisbon region.

In Portugal, there is neither a regional R&D policy nor a R&D regional policy. In fact, it is centrally formulated and implemented, especially with the strategic aim of attaining higher standards of scientific recognition. It is not surprising, that, in these circumstances, its profile is not very market-oriented. Being essentially fixed at national level, this public policy strengthens vertical hierarchical links and

Table 12.2 Portuguese S&T system: key indicators. Source: Eurostat (2015)

	Research and development expenditures by sectors of performance (% of GDP, 2013)	Gross domestic expenditure on R&D (GERD) by source of funds (% of total GERD Business enterprise sector, 2012)	Research and development personnel, by sectors of performance Head count (% of the labour force, all sectors, 2012)	Employment in high- and medium-high-technology manufacturing sectors (Share of total employment, %, 2014)	Patent applications to the European Patent Office (EPO) (number of applications per million inhabitants, 2014)	Human resources in science and technology as a share of labour force—Total (%, total, 2014)
Portugal	1.33	42.3	0.88	3.00	10.40	33.0
EU 28	2.03	48.1	1.12	5.70	111.74	44.4

Table 12.3 S&T indicators by NUTS 2. Source: Eurostat (2015)

	Human resources in science and technology (HRST), by NUTS 2 region (% of economically active population, 2015)	Employment in high-tech sectors (high-tech manufacturing and high-tech knowledge-intensive services), by NUTS 2 region (% of total employment, 2012)	Patent applications to the EPO by priority year, by NUTS 2 region (number of applications per million of inhabitants, 2012)	Total intramural R&D expenditure (GERD), by NUTS 2 region (% of GDP, 2013)	Researchers, all sectors, by NUTS 2 regions (% of total employment, 2012)
Norte	27.5	1.53	23.74	1.42	0.79
Centro	27.3	1.32	15.64	1.30	0.80
Lisboa	42.6	4.18	32.44	1.67	1.64
Alentejo	26.2	2.39	6.73	0.45	0.30
Algarve	24.6	2.10	1.50	0.37	0.40

centralism instead of acting as catalyst of territorially-based innovation dynamics. This way, the R&D policy in Portugal has been an instrument for accentuating growing disparities among the territories.

Nonetheless, it should also be remarked that the geographic dissemination of the R&D organizations, on the orbit of the universities of Porto, Minho, Aveiro and Coimbra, constitutes a solid enabling reason for promoting a regional innovation strategies. The localization of the research infrastructure between university and industry shows a noteworthy concentration in the more developed and higher density territories of North and Centre regions and should be considered a 'plus' for the formulation of regional innovation strategies.

12.3.2 The Mismatch Between the Knowledge Production Sphere and the Economic Sphere

The overall regional innovation system is defined by an unquestionable gap between knowledge production, namely the S&T system, and the productive sector. The S&T infrastructure has been acting according to an endogenous logic and, in doing so, does not match the entrepreneurial evolving demand. On the other hand, a vast number of SME entrepreneurs have low-level educational profiles. Typically, companies do not have enough qualified human resources to establish dialogue channels with universities and research centers. This situation combined with the preponderance of traditional and low-tech industries that still rely on scale and volume strategies rather than on innovation and differentiation results in a weakly structured demand-pull. The vast majority of SMEs habitually require a knowledge that is frequently under the codified S&T expertise of the academia.

Three programming periods of the co-funded EU support, already comprising competitiveness and innovation objectives, have only produced superficial organizational outputs in targeted Objective 1 territories (Figueiredo 2007), failing to form a closer cooperation amongst regional innovation agents.

In an effort to minimize the fissure between academia and the productive sector, some innovation-related agencies were launched, such as the Innovation Agency. However, these interface organisations mainly belong to the national innovation system which has a vertical and highly hierarchical orientation that hinders the creation of horizontal co-operative comportments among the territorial actors, stifling any possible synergies.

12.3.3 A Misconception of Innovation

Above all, companies adopt a competitive position based mainly on incremental and tangible products and process innovations where the top strategic priority is to improve production processes, productivity levels, and logistic channels, whilst decreasing labour intensity.

Enterprises have, in general, been adopting *fordist* strategies based on the search for decreased product prices. Tangible process innovations are the real *leitmotiv* of their market approach and positioning while other, more intangible categories of innovation—such as organizational and commercial innovations—fulfill only a minor role. This is a direct effect of the misconception of innovation amongst a large number of entrepreneurs as they assimilate modernization, founded on the renovation of capital goods with innovation.

So, there is an increased awareness about the need to change the basis for the competitive advantage of Portuguese less favored regions. RD&I competences focused on the integration of strategic are still new, lacking capillarity to be disseminated and absorbed by the traditional industries that vertebrate low density peripheral areas of Portugal (Simões 2008; Simões and Santos 2008).

12.3.4 A Deficit of Regionally Embedded Innovation Networks

Usually, the most important corporate and institutional partners alongside the value chain are not situated in peripheral territories. Subsequently, the innovative effort is not regionally embedded and it does not contribute for the densification of the territorial networking. As a result, firms remain uninformed of the local and regional updated knowledge transfer flows. Thus, it is difficult, in this context, to affirm that there are dynamic and aware territorial innovation systems, since they are virtually non-existent at a regional scale and also because the national innovation system is nearly absent from the genuine necessities of this set of enterprises. Codified S&T knowledge is shared through informal regionally-based networks, in which information circulates and is socialized. The firms' partners along the value chain are usually not in the regions and the innovation dynamics is not regionally embedded. The vast majority of the productive fabric seldom establishes other links outside the commercial partners of suppliers and clients.

So, besides their dimensional handicap, as the vast majority of the Portuguese enterprises are small to medium-sizes, the critical blockage is their (self-)segregation, not to be associated to the multi-channel flows, to the global world, the so-called loneliness syndrome.

In a convergent way, what might be called a collective learning process is not institutionalized because although an entrepreneurial culture exists that is based on empirical knowledge accumulated over generations, companies and institutional actors ultimately follow individualistic paths that do not enrich the local and regional environments in which they operate—in other words, it is not regionally established what might be called a true culture of contact.

In practice, regional innovation systems in Portugal are therefore non-existent or, not being so distrustful, embryonic. There are entrepreneurial and institutional stakeholders, there is institutional thickness (Amin and Thrift 1994), but there is a lack of a strategic collective dynamics, thickness is not converted into capability.

12.4 Redesigning Public Policies Conducive to Innovation

The focus of this section is on the problems faced by peripheral regions in overcoming comparative disadvantages in regards to their innovative capacity, as well as the public policies that can be promoted to reduce these handicaps.

Until about three decades ago, innovation policy in peripheral areas was often only understood as a supply-side dysfunction, in accordance with the prevailing paradigm of the linear model of innovation. Government policies, according to this framework, were usually designed to support the production of knowledge, for example, by providing incentives for R&D activities.

Garmise and Rees (1997: 2) state that "for the less favored areas of Europe and elsewhere, the relative lack of economic dynamism is rooted in very limited learning abilities of their innovation systems." The main focus of public intervention in this field should therefore be oriented in promoting processes for interactive learning involving all the different regional actors.

Corroborating this statement, Morgan (1997: 501) adds: "I would suggest that this is precisely what innovate on peripheral regions means, working with what exists, by inauspicious it may be, or appear, in an effort to break the traditional institutional inertia in the public and private sectors, fostering inter-networks that engage in collective process of interactive learning, cementing confidence capital."

Thus, a regional differentiation strategy becomes crucial to make better use of these specific territorial resources, for example, the existing cognitive stock, which is to serve as a baseline for new paths of upgrading and diversification, or even another perspective, the technology transfer system, which should be improved with regard to the specific needs of low-tech SMEs, since often the profile of demand for S&T of factors is not adequately answered by the traditional technology and knowledge transfer institutions (Tödtling and Trippl 2005; Hauser et al. 2007; Prange 2008).

In this format and content, the current innovation policy, as shown on Table 12.4, stresses the urgency of adaptation to different territorial idiosyncrasies. It also gets closer to the characteristic approach of modern regional policies that puts the focus on collective learning processes and institutional innovation instead of almost exclusively on the provision of infrastructure (Henderson and Morgan 1999: 19), and in attracting international mobile investment. This approach works to address the causes, not simply the symptoms of structural "backwardness" (as they were traditionally termed) of some territorial spaces. In fact, core of the strategy relies, to a large extent, on fighting the innovation gap that is characteristic of peripheral and structurally weak regions. In this sense, it can be said that this innovation policy, evolving from S&T policies, incorporates an increasingly important regional dimension and encompasses the promotion of modern innovation dynamics. Moreover, at the operational level, and even at the level of the respective conceptual framework, there is a notorious approaching trend, and even sometimes fusion, between these two *twin* policies that value above all the so-called development *software*, electing, as intervention priorities, the cognitive, intangible, organizational and institutional dimensions (Maillat 1998; Evangelista et al. 2002; Shearmur 2011).

Table 12.4 From science policy to innovation policy—a synthesis. Source: Adapted from Santos (2003)

	Science policy	Science and technology policy	Technology policy	Innovation policy
Period	Post 1st world war till the 1960s	From the 1960s to the 1970s	From the 1970s to the 1980s	End of the 1980s/beginning of the 1990s
Conceptual framework	*Science-push* model; logics of the activities dictated by the scientific community	*Science-push* model; growing attention to the information diffusion and transfer mechanisms	*Science-push* and *demand-pull* models; Increasing concern with the applicability of the scientific knowledge	Innovation interactive model; innovation systems; new paradigms of territorial development (Milieux Innovateurs, intelligent regions, regional innovation systems); institutionalization of the EU innovation policy
Objectives	To increase the knowledge levels; to promote the growth of the driving industries of the 2nd industrial revolution.	To increment the projection of the scientific research on socio and economic development; to support the constitution of firms with international dimension.	Absolute priority to technological applications; Modernization of the business world based on new technologies of information and communication; special focus on high-tech sectors.	Promotion of collective learning mechanisms; Creating positive agglomeration externalities; Adjustment of the dynamics of supply and demand of innovation factors; Strengthening endogenous innovation dynamics; Fertilization of industrial know-how with tertiary know-how; Coverage of high, medium and low-tech sectors.
Instruments	Promotion of public R&D activities; Creation of laboratories and academic science centers	Implementation of sectorial technological centers, technical and professional centers and information and dissemination platforms	Creation of S&T parks and of technopoles; Reconfiguration of laboratories and R&D centers in order to favor the production of applied technology; Tax and financial incentives for the adoption of best-practices	Promotion of business and institutional cooperation networks; Establishment of intermediation platforms between the business and the S&T actors; Regionalization of the research function; Tax and financial incentives to promote innovation entrepreneurial potential; Establishment of auditors' networks especially dedicated to the objectification of the technological needs of SMEs

In this sense, territorial revitalization policies cannot simply be distributive or *end of the line* repairing tools to minimize the crises affecting these regions. It is important that this new territorial development configuration does not narrow the material base on which the logic of local and regional development is anchored. They should instead be broad spectrum policies that, without losing the strategic aim of their interventions, may produce a mix that allows territories to reposition as protagonists of their own future: they should be understood as open to the world and accepting the inevitability of globalization either as a threaten and, mainly, as an opportunity; and they also must be understood as promoters of all initiatives that represent local and endogenous dimensions, focusing on territorial differentiation strategies (Santos and Caseiro 2015).

There is an understanding that the structural handicaps and constraints of the innovation dynamics in peripheral regions are generally less associated with the production of strategic information and knowledge and more related to processes that influence their dissemination and absorption by regional actors (Santos 2000). In these circumstances, it is crucial to provide aid mechanisms to minimize or solve these structural bottlenecks and to create opportunities for these regions to use strategic information to support innovation. It is also increasingly a false evidence, as argued by Veltz (1996: 194), "the idea of a technological progress, exogenous to the economic universe, that presents itself as a quasi-public good." This new generation of policies was developed try to address this broad spectrum of business actors who had not yet properly perceived the need to base the respective competitive strategies on innovation as a differentiating factor. The intervention focus is then strategically put on SMEs alert to innovation dimensions, trying to promote a set of technological and organizational externalities that can be absorbed by these companies based on an approach *from below*, as suggested by Capello (2014).

Henderson and Morgan (1999), call this new generation of territorial policy of *regional experimentalism*, thus encouraging exploratory dimensions and learning opportunities (*learning by experimenting*), seeing it mainly as a tool to develop social capital amongst various stakeholders involved—from the establishment of permanent channels of dialogue, the implementation of common projects that lead to the strengthening of trust and reciprocity ties, the growing interaction between the public and private sectors, the implementing institutions with functions brokering (bridging Initiatives), particularly in the field of entrepreneurially relevant information and knowledge transfer and incubation of innovative companies, promoting a network of supply of strategic business support services specifically targeted to the real needs of the productive sector, etc.—Maillat (1998: 16) argues this strategy is nothing but an attempt to play with the effect of territorial proximity, coupling industrial and tertiary knowledge.

Innovation policy understood this way abandons the casuistic attempts to promote and enhance technology transfer channels and stimulate the regional *milieu*. The core question, then, is whether less prosperous regions that have production bases considered less innovative and competitive at the international level can meet the necessary conditions to upgrade their social and cognitive capital. Landabaso's

(2003:16) cautious words are an important consideration here—they warn that it is necessary to adapt the innovation policy strategies at the various territorial contexts "as the innovation process does not follow the principles of chemistry: the mixture in each region required to produce a "reaction" (that is, to maximize the impact of innovation in the development) is different."

The most successful regions are those which are characterized by the ability of firms and institutions for adopting voluntary learning dynamics—in products, processes and organizational structures—and to adapt to the pressures induced by market dynamics (Henderson 2000; Santos 2009; Camagni 2014). Political intervention emphasis should, accordingly, move from the enterprise level (micro) to the level of the *milieu* itself (meso), since it is assumed that it is precisely the innovative territorially embedded dynamics, not necessarily each firm taken individually, which is responsible for the regional innovation upgrading process (Table 12.5). This assumption has implicit the recognition of the importance of externalities in the processes of innovation and diffusion, which seems justification enough to undertake public intervention, without which firms, especially SMEs, cannot fully develop all their innovation potentials. Thus, this constitutes its added-value in relation to the traditional industrial policies.

Following this line of reasoning, Pires et al. (2000: 1) importantly note that "innovation policies must have the fundamental mission of promoting the competitiveness of the productive system in a context of globalization of economic relations and the acquisition of competitive advantages resulting from the ability to innovate." In peripheral and depressed economies, innovation policy faces a double challenge: on one hand, upgrading the competitive profile of the companies associated with the most representative sectors of the different industrialization models of those territories and, on the other hand, of contributing to the emergence of new vectors of productive specialization, trying linkages to new and more demanding activities in

Table 12.5 Traditional industrial policies *versus* innovation policies. Source: Adapted from OECD (2011)

Traditional industrial policies	Modern innovation policies
Knowledge understood as a public good	Institutional and entrepreneurial empowerment as a learning process
Focus on technological innovation (product and process)	Broad spectrum of innovative production (hard and soft, including also organizational, market and social innovations)
Focus on high-tech firms	Inclusive logics, encompassing medium and low-tech firms and traditional sectors
Based on R&D institutions	Strategically focused on firms and, mainly, on the socioeconomic *milieu*
Knowledge diffusion as the main instrument	Stimulation of the absorption capability of firms and on networking promotion
Competitive advantages	Differential advantages; built advantages

S&T inputs, and also to provide a real increase of technical knowledge (OECD 2011; Santos 2012; McCann and Ortega-Argilés 2016). It thus includes a development framework that may help to diversify the economic profile of those territories, which are often too narrow and fragile.

One of the chronic handicaps that typifies these regions is related to the fact that their technological patterns are characterized by a S&T system in the public sector (universities, R&D laboratories, etc.) that is over-represented relative to the effort developed by the private sector. This implies normally consequences on the direction of research activities that are carried out that in these contexts, guided mainly by internal academic logic, more directed to stages upstream, towards focusing on the fundamental and applied research, moving away from the market needs (Santos 2003; Koschatzky 2003).

The creation of bridging platforms between the academic and the business communities, triple helix cooperative tools, as well as the regionalization of research, are strategies that aim to stimulate the local and regional *milieux* by building up sustainable territorial innovation systems (Figueiredo 2007; Cooke 2008; Santos and Simões 2014).

In this context, the instrumental focus should avoid the multiplication of wrong strategies, as argued by Landabaso (2003). In the recent past, public funds were injected in less-developed regions in an effort to promote the entry of more 'science' in the S&T system, which, by the fact that the latter is disconnected from the profile of the industrial fabric, could not find a translation into increasing regional innovation outputs, not reflecting that effort on corporate and territorial competitiveness. This is what Landabaso warned against.

In remote areas with fragile economic structures, the reorganization of traditional industrial sectors clearly constitutes one of the main challenges that innovation policy needs to equate. In this case, the establishment of mechanisms leading to the externalization of the firms RD&I must be considered (Lagendijk 2011). Moreover, one of the core problems that is crucial to attack is related to the fact that these depressed territorial areas are typically affected by very limited learning abilities that are the real cause of their economic anemia and, accordingly, the main focus of intervention public should be based on the promotion of enlarged, inclusive and collective learning dynamics, of catching-up and of institutional reorganization (Cooke 2007; Simões et al. 2008; Laranja 2009).

This new set of policy instruments, requires a high degree of decentralization in their design, delivery and management, as well as a consensual and cooperative work among the various actors involved. They should not being founded on large scales or in infrastructural projects and should consider the need for clear, territorial leadership without which it seems difficult to bring together the various rationales into play (OECD 2013). This suggests, of course, that an increase in regional capability for innovation inevitably involves new forms of organization and institutional partnership to help improve the structural competitiveness of the companies (Cooke et al. 2005; Santos and Simões 2014). If we agree that the intervention by the authorities should give priority to the implementation and strengthening of a relational culture, then policies have to comply with the existing overall network architecture and its specific territorial assets rather than focus more on punctual and atomized actions (Borrás and Edquist 2013). The strategic aim directed towards the reinforcement of the mechanisms for horizontal coordination and partnership, as well as interface management, avoiding political intervention supported in sectoral logics or fragmented actions.

It is in this light that is worth quoting the words of Morgan (1997: 501) when he says that "to innovate in peripheral areas means precisely work with what exists, even if it is not especially auspicious, in an effort to break the traditional institutional and corporate inertia, promoting inter-cooperation networks, involving actors in a dynamic of shared learning and feeding the reciprocal relations of trust." This is captures the greatest challenge territorial innovation policy faces in peripheral areas with structural development problems.

12.5 Conclusion

Today, there is a general acknowledgement that previous efforts increase investment in peripheral regions with structural development problems have not succeeded. Sharper, more wide-ranging approaches to creating territorial competitiveness are necessary. It is important to consider that integrating less-developed regions into the global knowledge economy has not yet been at a priority of regional development policies.

Territorial competitiveness has become a serious concern, mostly now that the pace of structural adjustment induced by the global economy is imposing new profiles of regional performance. Thus, the promotion of territorially embedded innovation systems appears to be a major challenge. For structural territorial competitiveness, sustainability has less to do with cost-efficiency and more to do with the capability to adapt to global dynamics and assume innovative strategies.

In peripheral territories, more attention should be put on the formulation and implementation of territorial innovation policies, and these policies should avoid the traditional sectoral supply-side approach (from above) as well as the orthodox instruments of some innovation policies. Innovation-led territorial policies should focus on catalyzing the relational and cooperative culture and on the mobilization of key stakeholders in collective learning dynamics.

The core of territorial innovation policies should concentrate on the stimulation of the whole regional *milieu*. In this way, a territorially embedded innovation policy should constitute a means of establishing a learning framework for all partners involved in the construction of a collective socio-economic trajectory. The remaining, fundamental question is how territories can organize themselves to further enhance policy innovations. This is the main challenge for many peripheral regions and a critical assessment must be done to determine the best course of action.

References

Almeida, J. (1994). *Sistema Regional de Inovação e Competitividade Industrial - o Caso da Península de Setúbal*. Lisboa: Instituto Superior de Ciências do Trabalho e da Empresa.

Amin, A., & Thrift, N. (Eds.). (1994). *Globalization, institutions and regional development*. Oxford: Oxford University Press.

Asheim, B. T., & Coenen, L. (2005). Knowledge bases and regional innovation systems: Comparing Nordic clusters. *Research Policy, 34*, 1173–1190.

Asheim, B. T., Smith, H. L., & Oughton, C. (2011). Regional innovation systems: Theory, empirics and policy. *Regional Studies, 45*(7), 875–891.

Bair, J. (2008). Analysing global economic organization: Embedded networks and global chains compared. *Economy and Society, 37*(3), 339–364.

Balland, P.-A., Boschma, R., & Frenken, K. (2015). Proximity and innovation: From statics to dynamics. *Regional Studies, 49*, 907–920.

Borrás, S., & Edquist, C. (2013). The choice of innovation policy instruments. *Technological Forecasting and Social Change, 80*, 1513–1522.

Camagni, R. (2014). The regional policy debate: A territorial, place-based and proximity approach. In A. Torre & F. Wallet (Eds.), *Regional development and proximity relations*. Cheltenham: Elgar.

Capello, R. (2014). Proximity and regional innovation processes: Is there space for more reflections? In A. Torre & F. Wallet (Eds.), *Regional development and proximity relation*. Cheltenham: Elgar.

Cappello, R. (1996). Industrial enterprises and economic space: The network paradigma. *European Planning Studies, 4*(4), 485–498.

Carlsson, B. (2005). Innovation systems: A survey of the literature from a Schumpeterian perspective. In H. Harmusch & A. Pyka (Eds.), *The companion to neo-Schumpeterian economics*. Cheltenham: Elgar.

CEC. (1997). *REGIS – Fourth interim report to EC-DG XII: Report on face to face interviews with regional firms – Aveiro region*. Aveiro: University of Aveiro.

Cooke, P. (1996). Building a twenty-first century regional economy in Emilia-Romagna. *European Planning Studies, 4*(1), 53–62.

Cooke, P. (2007). To construct regional advantage from innovation systems first build policy platforms. *European Planning Studies, 15*, 179–194.

Cooke, P. (2008). Regional innovation systems: Origin of the species. *International Journal of Technological Learning, Innovation and Development, 1*(3), 393–409.

Cooke, P., Clifton, N., & Oleaga, M. (2005). Social capital, firm embeddedness and regional development. *Regional Studies, 39*, 1065–1078.

Eurostat. (2015). *Eurostat regional yearbook 2015*. Luxembourg: Publications Office of the European Union.

Evangelista, R., Iammarino, S., Mastrostefano, V., & Silvani, A. (2002). Looking for regional systems of innovation: Evidence from the Italian innovation survey. *Regional Studies, 36*(2), 173–186.

Ferrão, J. (1997). Meios inovadores em cidades de média dimensão: uma utopia razoável? O caso de Évora. In J. Ferrão (Ed.), *Políticas de Inovação e Desenvolvimento Regional e Local*. Lisboa: Instituto de Ciências Sociais da Universidade de Lisboa.

Figueiredo, A. (2007). *Regional innovation systems as policy tools in knowledge oriented cohesion policies – The case of Portugal*. Paper presented at the Regional Studies Association International Conference. Lisbon: Regions in Focus.

Fujita, M., & Krugman, P. (2004). The new economic geography: Past, present and the future. *Papers in Regional Science, 83*(1), 139–164.

Garmise, S., & Rees, G. (1997). The role of institutional networks in local economic development – a new model of governance. *The Journal of the Local Economy Policy Unit, 12*(2), 43–56.

Hauser, C., Tappeiner, G., & Walde, J. (2007). The learning region: The impact of social capital and weak ties on innovation. *Regional Studies, 41*, 75–88.

Henderson, D. (2000). EU regional innovation strategies: Regional experimentalism in practice? *European Urban and Regional Studies, 7*(4), 347–358.

Henderson, D., & Morgan, K. (1999). *Regions as laboratories. The rise of regional experimentalism in Europe*. Cardiff: Cardiff University.

Keeble, D., & Wilkinson, F. (1999). Collective learning and knowledge development in the evolution of regional clusters of high technology SMEs in Europe. *Regional Studies, 33*(4), 295–303.

Koschatzky, K. (2003). The regionalization of innovation policy: New options for regional change? In G. Fuchs & P. Shapira (Eds.), *Rethinking regional innovation: Path dependency or regional breakthrough?* London: Kluwer.

Lagendijk, A. (2011). Regional innovation theory between theory and practice. In B. Asheim, R. Boschma, & P. Cooke (Eds.), *Handbook of regional innovation and growth*. Cheltenham: Edward Elgar.

Landabaso, M. (2003). *Clusters and less favored regions: Policy options in planning and implementation*. Paper presented at the UK Regional Innovation Network (RINET) Meeting. Stratford-Upon-Avon.

Laranja, M. (2009). The development of technology infrastructure in Portugal and the need to pull innovation using proactive intermediation policies. *Technovation, 29*(1), 23–34.

Lawson, C., & Lorenz, E. (1999). Collective learning, tacit knowledge and regional innovative capacity. *Regional Studies, 33*(4), 305–317.

Lundvall, B. A. (Ed.). (1992). *National systems of innovation: Towards a theory of innovation and interactive learning*. London: Pinter.

Maillat, D. (1998). Interactions between urban systems and localized productive systems: An approach to endogenous regional development in terms of innovative milieu. *European Planning Studies, 6*(2), 118–130.

Maskell, P., & Malmberg, A. (1999). The competitiveness of firms and regions: "Ubiquitification" and the importance of localized learning. *European Urban and Regional Studies, 6*(1), 9–25.

McCann, P., & Ortega-Argilés, R. (2016). Smart specialisation, entrepreneurship and SMEs: Issues and challenges for a results-oriented EU regional policy. *Small Business Economics, 46*(4), 537–552.

Morgan, K. (1997). The learning region: Institutions, innovation and regional renewal. *Regional Studies, 31*(5), 491–503.

Morrison, A. (2008). Gatekeepers of knowledge within industrial districts: Who they are, how they interact. *Regional Studies, 42*(6), 817–835.

Mota Campos, M. I. & Silva, M. R. (1997). *Política de inovação em regiões industriais atrasadas*. Paper presented at the IV Encontro Nacional da APDR. Covilhã: Universidade da Beira Interior.

Moulaert, F., & Sekia, F. (2003). Territorial innovation models: A critical survey. *Regional Studies, 37*(3), 289–302.

Nicolau, I. (2001). Trajectórias de inovação e ciclo de vida da indústria - O distrito industrial de Alcanena. In C. Antonelli & J. Ferrão (Eds.), *Comunicação, Conhecimento Colectivo e Inovação - as Vantagens da Aglomeração Geográfica*. Lisboa: Instituto de Ciências Sociais.

OECD. (2011). *Regions and innovation policy*. Paris: OECD Publishing.

OECD. (2013). *OECD reviews of innovation policy: Sweden 2012*. Paris: OECD Publishing.

Pires, A. R., Rodrigues, C., & Castro, Eduardo e Esteves, Carlos. (2000). *Construção de uma Estratégia de Inovação para a Região Centro*. Aveiro: CEIDET-UA.

Prange, H. (2008). Explaining varieties of regional innovation policies in Europe. *European Urban and Regional Studies, 15*, 39–52.

Putnam, R. D. (1993). The prosperous community – Social capital and public life. *The Americam Prospect, 13*, 35–42.

Santos, D. (2000). Innovation and territory: Which strategies to promote regional innovation systems in Portugal? *European Urban and Regional Studies, 7*(2), 147–156.

Santos, D. (2003). Política de inovação: filiação histórica e relação com as políticas de desenvolvimento territorial. *Revista Portuguesa de Estudos Regionais, 3*, 25–40.

Santos, D. (2009). Teorias de inovação de base territorial. In J. S. Costa & P. Nijkamp (Eds.), *Compêndio de Economia Regional - Teoria, Temáticas e Políticas*. Cascais: Principia.

Santos, D. (2012). *Dinâmicas Territoriais de Inovação no Arco Urbano do Centro Interior: O Caso do Setor Têxtil-Confeções*. V.N. Famalicão: Húmus.

Santos, D., & Caseiro, N. (2015). The challenges of smart specialization strategies and the role of entrepreneurial universities: A new competitive paradigm. In L. Farinha, J. Ferreira, H. Lawton-Smith, & S. Bagchi-Sen (Eds.), *Handbook of research on global competitive advantage through innovation and entrepreneurship*. Hershey: IGI Global.

Santos, D., & Simões, M. J. (Eds.). (2008). *Dinâmica Socioeconómica da Fileira da Madeira em Concelhos do Pinhal Interior Sul: uma Análise através do Conceito de Meio Inovador*. Covilhã: Universidade da Beira Interior.

Santos, D., & Simões, M. J. (2014). Regional innovation systems in Portugal: A critical evaluation. *Investigaciones Regionales, 28*, 37–56.

Shearmur, R. (2011). Innovation, regions and proximity: From neo-regionalism to spatal analysis. *Regional Studies, 45*(9), 1225–1243.

Simões, M. J. (Ed.). (2008). *Dos Projectos às Regiões Digitais: Que Desafios?* Lisboa: Celta.

Simões, M. J., & Santos, D. (2008). Challenges in the digital cities and regions in Portugal. In P. Cunningham & M. Cunningham (Eds.), *Collaboration and knowledge economy – Issues, applications, case studies*. Amsterdam: IOS Press.

Storper, M. (1995). The resurgence of regional economies, ten years later: The region as a nexus of untraded interdependencies. *European Urban and Regional Studies, 2*, 191–221.

Storper, M., & Scott, A. J. (1995). The wealth of regions. Market forces and policy imperatives in local and global contexto. *Futures, 27*(5), 505–526.

Tödtling, F., & Trippl, M. (2005). One size fits all? Towards a differentiated regional innovation policy approach. *Research Policy, 34*, 1203–1219.

Torre, A., & Wallet, F. (2014). Introduction. The role of proximity relations in regional and territorial development processes. In A. Torre & F. Wallet (Eds.), *2014, regional development and proximity relations, new horizons in regional science*. London: Edward Elgar.

Tura, T., & Harmaakorpi, V. (2005). Social capital in building regional innovative capability. *Regional Studies, 39*, 1111–1125.

Veltz, P. (1996). *Mondialisation, Villes et Territoires. L'Économie de l'Archipel*. Paris: PUF.

Chapter 13
Conclusion: Resilience—What's Next?

Eric Vaz, Teresa de Noronha, and Hugo Pinto

The previous chapters of this book did not explore the concept of resilience from the perspective of firms. Corporations, including small and medium-sized firms, are those to function as the real engines of the productive processes. They have been under extreme cost reduction policies due to great instability related to both medium-term severe technological change and numerous volatilities expected within the financial markets. To maximize potential cost reduction and mitigate loss, companies are obliged to add flexibility to their whole production process using salary reduction. It is predicted that continuous excessive downsizing will no longer be a sustainable solution for these issues.

Successful companies require excellent skills to provide adequate solutions. Simultaneously, powerful technological tools distribute and supply the productive chain with choices for long-distance networking, learning, buying, distributing and monitoring at global scale—choices that must be made available to mitigate costs. New opportunities spread over the global market are likely to significantly alter the concept of space and time, including the advantages of agglomeration economies. Further, the value of scale prevails yet over the value of space, with the decreasing costs of distribution and transportation occurring as technology advances.

For now, the general difficulties of firms dealing with fast-technological changes within the supply chain are still an obstacle and delay the disruptive process behind the classical concept of space and time economies. However, such barriers will soon be overcome by the increasing inclusion of information and communication technologies in corporative management, public services skills, and strategies altogether.

E. Vaz (✉)
Department of Geography and Environmental Studies, Ryerson University, Toronto, ON, Canada
e-mail: evaz@ryerson.ca

T. de Noronha · H. Pinto
Faculty of Economics, University of the Algarve, Faro, Portugal
e-mail: mtvaz@ualg.pt; hpinto@ualg.pt

H. Pinto et al. (eds.), *Resilience and Regional Dynamics*, Advances in Spatial Science, https://doi.org/10.1007/978-3-319-95135-5_13

Based on an extensive set of studies by the authors (Noronha Vaz and Nijkamp 2009a, b) concerning the evolution of regional innovation patterns as well as systems versus town size, our first conclusion is that a new emphasis in the theoretical framing of scale versus agglomeration economies that are affecting directly the resilience of a territory. Resilience is the fine-tuning balance between the progress or decay of organizations in a context of increasing or stagnant territorial opportunities.

Maintaining resilience means being resistant to change while accepting that, from a socio-economic perspective, resistance holds positive and negative aspects. As described in several chapters of this book, the desire is that resistance will only occur when obstacles arise. However, we must be alert for a concept of a hidden inelasticity of the productive structure that may be able to affect situations of progress as well. This directly leads to our second conclusion, which emphasizes some critical aspects of the process itself, where we find ourselves seeking for factors able to strength the territory for losses of unpredictable nature. Our third conclusion notes that there is no short-cut to implement resilience without integrating innovation, technological innovation, and the forms how it may protect the entrepreneurial ecosystems and their social contexts—knowledge-based skills and respective mobility.

All the three conclusions above mentioned asking for public policy support measures driven by intelligent and diligent strategies focusing on those factors that have contributed to socio-economic progress in a context of instability. Thus, assessing the behavior of organizations in fragile ecosystems is mandatory.

Based on the understanding of local innovation and entrepreneurial capacity of regions and locals, we can hence suggest that the causal forces and socio-economic impacts of climate change in a context of sustainability are critical. In a practical case, we suggest that regional or local imbalances must be studied to develop a better understanding of the potential of regional companies to dynamically innovate. Therefore, quantitatively discussing and analyzing patterns of innovation in companies, using them as facilitators of resilient territories, and observing their temporal advances in creating links and networks is necessary.

This book provides a framework for the analytics of resilience at four different levels:

The Technological Development Paths Far beyond the cognitive issue, the complexity of the innovation system is structured in conditions relating to governance systems and respective space-time industrial organization. This argument appeals to the interpretation of Schumpeter on the propensity of innovators to group geographically and can generate groups, using innovation not only as a dynamic element to the company but also as a powerful instrument of growth of the environment. On this basis, innovation and their factors have become of critical concern, and the complexity of governance systems is one of the key vectors to explain the success of efforts to promote innovation and sustainability. There are numerous ways to identify these factors: Some authors adopted a more sustained vision in the company's resources, taking the heterogeneous character of the companies and emphasizing its strategic behaviour, for example (Noronha and Cesário 2008).

The Networking Systems With the recognition of knowledge as a key resource for companies and other economic agents, there is evidence that relations between industry and external research organizations to companies for successful transfer of technological knowledge is essential. This is a concept that has been successively stretched and is referred to as the "Triple Helix" concept. The emphasis on a triangular interaction between research communities, governments and industries and denotes a joint effort of territorial scope. With durable and consistent inter-institutional links, it is possible to be observed consequent settings in forms of networks and industrial clusters. Indeed, a variety of studies on clusters have become influential in describing how and why the institutions come together to respond to various kind of pressures.

Strategic Choices of Companies and the Spatial Impacts of Its Network Organization As indicated above, when viewed from a global perspective, the structures tend to outline the trends in long-term technology and can help explain the difficulties in reducing the growth of different capacities between countries and regions and/or places. In general, the diverse causes of this behavior, as well as a propensity to have cyclical disadvantages in many areas of the world later, have attracted the attention of many. The Italian School founded by Camagni (1991, 1995), later strengthened by researchers northern Europe, such as Asheim and Isaksen (2003), realizes a direct contribution of individual companies or industrial clusters to regional growth. More recently, such studies tend to discuss further regional or local resilience to stress. This was further emphasized in the long investigation related to the side effects "spill over effects" developed by Kaiser (2002) and Fischer (2006).

Regional Strategic Learning If the business environment is consistently framed with the presence of significant links, the clusters that can be identified, despite the uncertainty that always prevails in economic environments, new face future needs related to their resources or their customers, in a context of increasing fragmentation of the production process. Often, innovative companies accumulate knowledge through learning as a process to reduce uncertainty and not necessarily to achieve economies of scale. In this case, the best decision of the acquisition of knowledge that may involve the entrepreneur in strategic learning as an opportunity to become resilient. Empirical studies, where it is often stressed the local context in which companies develop their activities in interactive mode between the parties and the set, have proven that organizational learning and institutional networks combine to increase the performance of innovative companies. Occasionally, companies can find possible solutions in specific networks for technological learning or other forms of learning and, through external sources, can manage the interfaces that help combine technical know-how (sources of information and relations), helping to create solid instruments for regional or local growth and development.

Altogether, this book has pointed out three major fields of work delineated on some of the multiple aspects that resilience represents: The foundations expressed as definitions and related concepts; The diversified contexts in which the phenomenon may occur; And, finally, the variety of strategies promoting resilience.

After much research related to the Economics of Innovation, we understand the concept of resilience with some intellectual frivolity and, and have reduced it to a simple state. However, in this book we emphasized the complex analysis related to the multiple and surprising effects of causality when dealing with diverse socio-economic agents. Thus, as each territory is unique by its legacy, its identity, and its culture, including the knowledge base and governance system, our fourth conclusive note refers to the complexity of selected paths to achieve resilience.

Although three fields of work have been included in this publication, the text indirectly calls for a logical structure which we would like to synthesize in this conclusion:

Human and financial capital are major determinants of growth and stability thereby promoting regional competitiveness driven by a technological adjustment to modernity. At a different level, resilience can be acquired employing socio-economic flexibility, either at the level of the structure of employment or using openness to the exterior. If territories are supposed to be resilient, they should get their support from the exterior. This assistance should be either by diversified commercial exposure, or by the existence of established and trustful networking systems or other forms of cooperation.

However, from a corporate point of view, a more microeconomic perspective is necessary to closely observe specific conditions that make firms more resilient to adversity. This is a discussion that cannot escape us as it calls specifically to the territorial structure and what it has to offer. Almost as a thematic artifact, we are obliged to appeal to the urban structure and the role of city-regions, the great winners of technological and organizational advantages. Finally, from a more regulation viewpoint, it will be the social capital and the structure of governance the facilitators of resilience.

Noronha Vaz and Nijkamp (2009a, b) emphasized the fact that, wanted or not, "both large-scale agglomeration and regional economic specialization are a persistent and growing phenomenon..." conducting to externalities such as "knowledge spill overs and dependence on human relations, rules, and customs that enable firms to coordinate under conditions of uncertainty."

We hope that this book calls the attention for the importance of the topic and excites scientists to further work on these issues. As we prepared the volume and wrote the Introduction and Conclusion, it became clear that ours is a modest contribution to the initial steps of a much border field of studies to be promoted. Across the world, there are diverse capacities to respond to challenges that remand reliance. It would be erroneous to reduce the solutions to a "one size fits all" strategy—indeed, that would destined to fail. New contributions and definitions are required to consolidate resilience in order to advance the analytical methods surrounding this crucial topic in the future.

References

Asheim, B. T., & Isaksen, A. (2003). SMEs and the regional dimension of innovation. In B. T. Ascheim (Ed.), *Regional innovation policy for small-medium enterprises* (pp. 21–46). Cheltenham: Edward Elgar.

Camagni, R. (1991). Introduction: From the local milieu to innovation through corporation networks. In R. Camagni (Ed.), *Innovation networks: Spatial perspectives* (pp. 1–9). London: Belhaven Press.

Camagni, R. (1995). Espace et temps dans le concept de milieu innovateur. In A. Ralllet & A. Torre (Eds.), *Économie Industrielle et Économie Spatiale* (pp. 193–210). Paris: Economica.

Fischer, M. M. (2006). *Innovation, networks and knowledge spillovers*. Berlin: Springer.

Kaiser, U. (2002). Measuring knowledge spillovers in manufacturing and services: An empirical assessment of alternative approaches. *Research Policy, 31*(1), 125–144.

Noronha Vaz, T., & Cesário, M. (2008). Driving forces for innovation: are they measurable? *International Journal of Foresight and Innovation Policy, 4*(21), 1–2, 30–50.

Noronha Vaz, T., & Nijkamp, P. (2009a). Knowledge and innovation: The strings between global and local dimensions of sustainable growth. *Entrepreneurship and Regional Development, 21* (4), 441–455.

Noronha Vaz, T., & Nijkamp, P. (2009b). Multitasking in the rural world: Technological change and sustainability. *International Journal of Agricultural Resources, Governance and Ecology, 8* (2), 111–129.

The manufacturer's authorised representative in the EU is Springer Nature Customer Service Centre GmbH, Europaplatz 3, 69115 Heidelberg, Germany. If you have any concerns regarding our products, please contact ProductSafety@springernature.com

Printed and bound by CPI Group (UK) Ltd, Croydon, CR0 4YY
06/10/2026
02228973-0003